LEADING
IN THE AGE OF
DATA

YOUR GUIDE TO THE 7 FACTORS
OF TEAM EMPOWERMENT

BEN JONES

Founder and CEO, Data Literacy

DATA LITERACY
PRESS

ISBN: 978-1-960907-03-5 (hardcover)
ISBN: 978-1-960907-02-8 (paperback)
ISBN: 978-1-960907-04-2 (ebook)

Cover design and interior graphics designed by Alli Torban

Printed in the United States of America.

For Ellie and Valerie,
two leaders who have changed the data world for the better.

MORE BY BEN JONES

The Data Literacy Series

Data Literacy Fundamentals

Learning to See Data

Read, Write, Think Data

Other Books

ChatGPT Basics

The Introspective Entrepreneur

Avoiding Data Pitfalls

Communicating Data with Tableau

CONTENTS

This book is a companion
to the Data Literacy for Leaders
online course which can be found at
https://dataliteracy.com/data-literacy-for-leaders

PREFACE

Everything is changing, all over again. The reality is that things never stopped changing. But as I sit here and type the last two pages that I'll write before sending this book to production, I have a distinct feeling that major changes are now underway in technology and in society. I'm not alone in feeling this way. Some people around the world are filled with excitement, some are filled with terror, and others like me are filled with an odd mixture of both. The reason for these strong feelings is artificial intelligence, or AI, and a series of developments that is building upon the massive amounts of data that we have been creating at an accelerated pace over the past few decades.

I just looked back at the date that I first created this document: November 29, 2022. That was the day before OpenAI launched their AI chatbot, ChatGPT, to the world. To say that it has taken the world by storm is a dramatic understatement. Over 100 million ChatGPT accounts were created in less than two months' time, by far the fastest adoption of a new technology in human history.

The changes triggered by OpenAI's tool and its competitors have been so drastic, in fact, that I temporarily put this book on hold to start and publish another book, titled *ChatGPT Basics*. My goal was to educate people about the technology that has been sweeping through every discipline and every industry. The operating belief at my company, Data Literacy LLC, is that when it comes to harnessing a powerful new technology, an educated population is far better than an ignorant one.

After completing the writing and publishing of that impromptu book, I was eager to return to this one, because I realized that we don't just need an educated population, we need strong leaders to help us chart a course forward. There are many uncertainties and concerns about how AI will affect our world. In order to prevent some of the predicted evils of AI, we need effective, diverse, data-savvy leaders in all three sectors: the public sector, the private sector, and the voluntary sector. AI isn't going away, and its potential is extreme, both to help and to harm.

I have dedicated this book to two leaders in the data world whom I have been fortunate enough to know and to work with. The first, Ellie Fields, hired me over a decade ago to run the marketing team for Tableau Public, a free visual analytics platform that has been used by hundreds of thousands of people to date to tell the stories of our time with data. Ellie is the best boss I have ever had. She blends a deep care for the work being done with a sincere interest in the people doing the work. Her leadership changed my life, and it had a massive impact on the data world.

The second leader I'm dedicating this book to, Valerie Logan, is the founder and the mother of the Data Literacy Movement. Her pioneering work at Gartner helped both to uncover and to explain the biggest education gap of our time. And now her work at her own company, The Data Lodge, is helping organizations close that gap with extremely competent counsel and a highly engaged community. Valerie—like my wife, Becky, and I—had the drive and the gumption to walk away from a successful role at a great company to take on an even bigger leadership challenge as the driving force behind an entire movement. Valerie has become a trusted confidant to Becky and me, and a source of inspiration as we tackle the difficulties associated with starting and running a business.

Here's my hope: that this book encourages more people to do what Ellie and Valerie have done—to step forward and become great leaders in this present age of data. What we need are wise leaders who understand how to use data and how not to use it, and we need leaders who care enough about people and the planet to make sure that data is used for good. If I'm brutally honest with myself and with you, that's the only way that I'll want anything to do with data or technology in the decades ahead.

Ben Jones
Bellevue, Washington
May 17, 2023

INTRODUCTION

It's imperative in today's world that leaders of organizations of all types and sizes be data-savvy. This is true no matter what department or industry leaders find themselves in. You're a marketing manager at a consumer goods company? You need to be data-savvy. You're a vice president of human resources at a nonprofit? You need to be data-savvy, too. The same goes for a facilities operations director at a government agency, and a department chair at a community college.

Every one of these leadership positions, and countless more, involve the use of data to achieve the key goals and objectives of the team. In today's world, there's just no escaping data. As a leader, you can either try to ignore data, or you can embrace it, harness it, and put it to good use for you.

But what exactly does it mean for a leader to be "data-savvy?"

Well, here's what it *doesn't* mean: it doesn't mean that the leader is a subject matter expert in data analytics or data science. It doesn't mean that they spend all of their time working with raw data housed in cutting-edge data lakes. It doesn't mean that they use sophisticated software or programming languages to mine and explore data themselves.

Then what does it mean? What does a data-savvy leader *do*? A data-savvy leader works to create a collaborative environment in which their team can use data to achieve their goals and objectives while dramatically reducing or eliminating any negative side-effects associated with their use of data. The data-savvy leader knows how to set the stage for success, and they do what it takes to identify any barriers or hindrances, and then reduce them to a manageable level.

In essence, then, the team's overall "data climate" is the leader's responsibility. The outcome is owned by the entire team, but the buck stops with the leader. This is true whether the leader happens to be a manager, a director, or an executive.

The fact is that the climate can go two different ways. The climate can either be conducive to making effective use of data, or it can be detrimental to the same. That's not all, though. Just because a team can use data doesn't mean that the outcome will be positive and constructive. The outcome of the team's use of data can go two different ways, as well: the outcome can either be helpful, or it can be harmful.

The reality, of course, is that these two concepts—data climate and data outcomes—aren't binary at all. At any given time, a particular team's climate will propel them forward in certain ways, and it will hold them back in other ways. And the outcomes will involve some results that are helpful, and other results that are harmful. Sometimes a result that seems helpful at first can reveal its harmful side effects down the road. Other times it's hard to tell how it will all turn out.

So it can be a mixed bag, and it's difficult to avoid all possible unintended consequences. The bottom line is that the data-savvy leader sees and preserves the good, and they spot and root out the bad. The end result is that the data climate and the data outcomes amount to a fruitful harvest for the company, for society, and for the planet.

I have a window into what makes a great leader in this present age of data, and I'd like to invite you to peer through that window, too. Let me tell you about this window to which I'm referring. In January 2020, my company, Data Literacy LLC, launched a new offering called **The Data Literacy Score: A Team-Based Assessment**. It's a subjective assessment that lets leaders hear what their team thinks about their collective ability to make effective use of data. It has been used by leaders of companies, nonprofits, and government agencies all over the world.

It's one lens through which a leader can evaluate the results of their own leadership in the age of data. It's not a perfect lens by any means, but many leaders have found it to be very enlightening. Commonly identified strengths and weaknesses start to emerge, and "data pain points" become easy to spot. The assessment rounds out with a set of ideas and recommendations that help the leader chart a course forward. The goal? Higher levels of data literacy. A great leader in the age of data is one that can bring out this increase in data maturity.

Let me give you more details about this assessment. The individual team members who respond and complete the assessment collectively give the team an overall Data Literacy Score from 0 to 500. The score places the team into one of five maturity levels: Data Novice (0 to 299), Data Aspiring (300 to 349), Data Inclined (350 to 399), Data Focused (400 to 449), and Data Wise (450 to 500). These scores roughly align to quintiles: of the thousands of individuals who have completed the assessment to date, almost exactly 20% have fallen into each of the five categories.

The Data Literacy Score: Maturity Stages

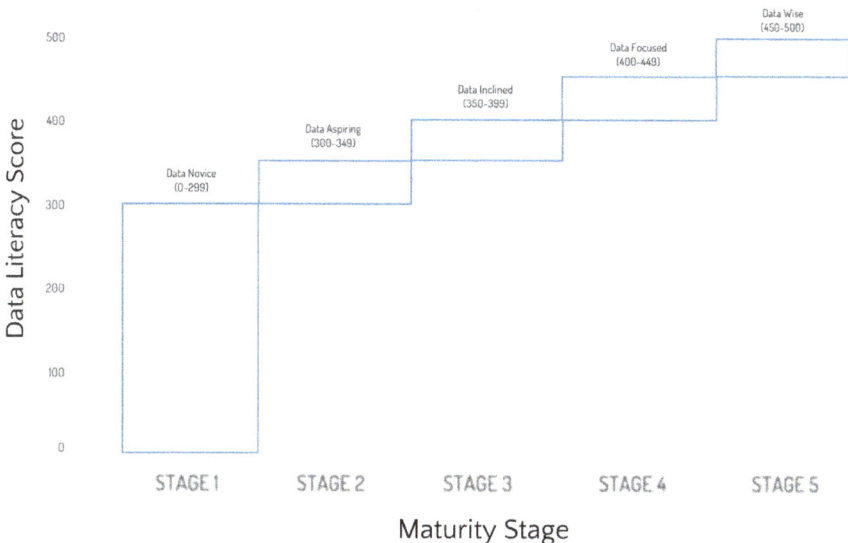

Where do these 500 possible points come from? The assessment features a bank of 50 scoring statements that are worded in a positive way: if the statement applies to your team, that would be a good thing. Respondents give each statement a score between 0 (does not apply at all to my team) to 10 (fully applies). Adding up the scores of the 50 statements, each individual respondent can therefore assign a total score between 0 and 500 overall. The team's overall score is simply the average of each individual respondent's total score. To give you a better idea of what these 50 scoring statements look like, here's one of them:

> *"It's clear that my team vales data by the way that we communicate with each other and by the way we work together with data to get things done."*

Does not apply at all	○	○	○	○	○	●	○	○	○	○	○	Fully applies
	0	1	2	3	4	5	6	7	8	9	10	

The 50 statements include a single, overarching statement along with 49 statements divided evenly into seven different categories: Ethics, Purpose, Data, Technology, People, Process, and Culture. The reason we included each of these seven categories is that we believe that a team's ability to effectively use data for good involves much more than how much data they have, and what software tools they have purchased. A good leader in the age of data needs to be aware of the health of each of these seven categories. Any one of them can be "make or break."

THE DATA LITERACY SCORE
A TEAM-BASED ASSESSMENT

PURPOSE
How well do we incorporate data into the value we provide to our various stakeholders?

ETHICS
Are we using data for good or for harm?

DATA
Are we making high-quality data available for our teams?

TECHNOLOGY
Do we have the right tools to leverage our data in a powerful way?

PEOPLE
Do we have employees with the necessary knowledge, skills, behaviors, and attitudes?

PROCESS
Do our processes allow for effective use of data, or do they get in the way?

CULTURE
What are we doing to enable, promote, & reward a data literate environment?

Let's briefly consider each of these seven categories one by one. Together they form the outline for this book: seven chapters that will challenge you to consider your approach to leadership as it relates to data.

Ethics

We start with the "true north" of every organization: it's the compass that points the way toward ethical behaviors and decisions. Ethics can be a tricky subject, and many people in business shy away from using the word altogether. I believe this is a mistake. As we have seen, when organizations turn a blind eye to the ethical ramifications of what they do, real harm can result.

Data presents organizations and their leaders with a long list of ethical challenges and dilemmas. From violations of privacy to propagation of societal biases to publication of inaccurate or misleading figures, when it comes to data, organizations and their leaders can get it wrong. There's no doubt about that. Effective leaders meet this challenge head on. In this first chapter, you'll consider seven different ways you can infuse sound ethical principles into your team's data practices.

Purpose

No one wants to work on something that doesn't matter whatsoever. We all want to be engaged in doing things that help us achieve a goal that we care about. In our lifetime, in this age of data in which we live, we have come to expect a lot out of our careers. Because of that, many of us want more than a paycheck from our employers; we want meaning.

That's difficult to bestow upon someone, and certainly each of your team members plays the most important role in finding meaning in their own life and career. But as a leader, you can make it crystal clear how their activities are translating into real value for those you are trying to serve. And you can make the linkage between data and value, too. What are your goals and how are you measuring your team's performance relative to them? What role does data play in achieving these goals? Don't let data be a pointless sideshow, used only in pet projects that don't move the needle.

Data

It's hard to talk about or even think about data literacy without focusing at some point on the data itself. If data is the most important and most valuable asset that your team has, after the team members themselves, then what kind of data are you asking them to work with? Is it sufficient to get the job done? Is it accurate enough and fresh enough? Does it have documentation or metadata to help your team members understand what it all means?

And let's talk about the balance between permissiveness and protectiveness. Can your team members actually get access to the data they had a legitimate need for? What about data privacy and data security? Are you taking care of sensitive data so that no one is harmed in the process of your team working with that data? These and many more questions are relevant for you to consider as your team's leader. Even if you're not in IT, you need to partner with others across the organization to ensure that your team's data is both rich and safe.

Technology

You can't make use of data today without interacting with multiple tools and technologies. Regardless of where in the data workflow your team spends the majority of their time, they'll need adequate tools to get the job done. From data collection to data storage to data preparation to analytics, visualization, and storytelling, the data toolbox is full of a wide variety of tools that your team will need to learn how to use. These tools are constantly evolving, and new ones come out every month.

Well, what technologies are you asking your team to use? Are they good enough? In this chapter, you'll consider how well you as a leader are equipping your team with data working tools. You'll look at their data tools from a variety of perspectives, such as the performance of those tools, whether they work well together, and whether they're flexible enough to adjust to the changing environment. Depending on your role and your level within the organization, you may or may not have a lot of control over the data working tools your team gets to use. Either way, though, it's your job to understand what tools are needed, and it's your job to advocate for them.

People

The most overused refrain by executives in every industry is that the organization's people are its most valuable asset. It can be hard to square that claim with the massive layoffs that happen in every industry from time to time. And it's easy to see how this phrase can be somewhat self-serving, because it's in the best interests of the executive that those employed by the organization feel highly valued so that they'll continue to be productive. So how do you as a leader cut through all the platitudes and lip service and make this phrase real? You do so by showing with your actions, not your words, that it's true.

If you highly value the people on your team, you'll invest in them. This is especially relevant to their level of data literacy. It's imperative that each person on your team achieves a basic level of data literacy. In order to grow in their careers, they'll need to advance beyond a basic level. In this chapter, you'll ask yourself what you are doing to give them the opportunity to grow and develop. Think of a plant that needs air, water, sunlight, and soil to grow. What training opportunities are you giving them so they can learn how to work with data? What job assignments are you giving them so that they can put this training to good use? What data-informed decisions are you exposing them to and even delegating to them? You can't realize the promise of data without data-savvy team members.

Process

The Ethics category is the most neglected category in this list of seven, but the most overlooked category is Process, and it's not even close. That's my perspective, because after over a decade working in business intelligence, I have heard or been involved in countless conversations about data and tools, but hardly any of those conversations touched on how those elements fit into a team's processes. A process is the set of steps or actions taken to achieve a particular end. Teams get work done via processes, whether they think of it that way or not. Some processes are rare or ad hoc. Others recur on a regular basis.

Either way, your team's processes either run smoothly or they do not. You and your team can achieve breakthrough results by marrying data and process. You can leverage data to make your processes more efficient. And you can design your processes to improve your data so that its value continues to grow. As your team's leader, you are the custodian of your team's processes. People on your team may be the ones carrying out the steps themselves, but you are the one responsible for the way the process is running. When your most important processes are cross-functional and require collaboration with other teams, who else will liaise with leaders across the organization?

Culture

The category with the highest potential to drive data literacy across your team and entire organization is Culture. What does "culture" even mean, though? Briefly, it's the values, norms, behaviors, and beliefs that a group of people share. Together, this collection of ideals and practices drive how people interact with one another inside the group, and they affect how group members interact with those outside of the group. If your group can be said to have a personality, then culture would be the driving force behind it.

Unless you're the CEO, your own team exists within the context of a broader organizational culture. As such, your team's culture will be heavily influenced by the overall culture. Even a CEO's team is influenced by higher cultures: those of the societies in which the organization operates. That's no excuse to abdicate responsibility for your team's culture, and great leaders find ways to shift, nudge, and nurture culture into the direction they want it to go. This is especially true of the way a leader uses culture to establish and reinforce the value of data. Leaders have constraints and limitations, but they also have a lot of power to clarify team purpose and vision relative to data, to make decisions about rewards and recognition for successes with data, and to create a positive environment of communication and collaboration around data. The final chapter, then, represents a major opportunity for you as a leader to set up your team for success.

When put together, these seven categories give you as your team's leader a massive amount of influence that you can use either to enable or to hinder data literacy across your team. Determining which of these categories to focus on and improve is up to you. In the coming chapters, you'll get a better idea of what each one entails so that you can assess their current state and their potential. I wish you all the best on your journey! I know this for sure: you have the ability to become a great leader in the age of data.

Chapter 1

ETHICS

"If ethics are poor at the top, that behavior is
copied down through the organization."

–Robert Noyce

When it comes to data and ethics, we really put the cart before the horse, didn't we? What do I mean? In what we can call "The Great Data Arms Race of the 21st Century," companies in every industry compiled data, deployed technologies, hired data talent, and implemented data-driven processes to profit and obtain competitive advantage as fast as they could.

Why wouldn't they? Many of these organizations are here for us to talk about them because they did just that. Many more have gone the way of the dodo bird because they didn't wake up to the need to evolve from whatever kind of company they were into a data company.

But in the mad frenzy to get a leg up in their respective marketplaces, many companies failed to take data ethics into account. They brushed aside important questions about the kind of harm that could potentially result from all that hoarding and leveraging of data: we're talking about dirty data, sensitive personal records, and algorithms that generate unfair outcomes.

"Just use it," they said. "It'll be fine," they said.

And what happened as a result? When you put the cart before the horse, what happens? Well, nothing, really. The cart and the horse just sit there, going nowhere. Unless they're on a slope. Then it could get ugly, or at least so we can imagine. The mental image of a runaway cart is an apt analogy, because that's exactly how data has behaved, and in many ways it has crashed right into society, and right into each one of us.

We've seen data breaches galore, invasions of privacy, and further entrenchment of biases against marginalized groups.[1] Students have been falsely accused of cheating on exams.[2] Artificial intelligence chatbots have turned hideously racist.[3] Automatic photo-tagging algorithms labeled Black people as gorillas.[4] Women have been served fewer ads related to high-paying jobs than men.[5] The list goes on and on.

The unavoidable fact is that very real damage has been done with data. Not just once, but over and over. This has led to increased defensiveness and polarization in our societies. It's actually not surprising at all that this is the case. When new technologies are first introduced, the resulting rate of harm can be quite high. As usage of these new technologies increases, we tend to find ways to prevent the kinds of damage that were widespread at first.

Take the example of the actual cart and horse. When this popular mode of transportation first began to be replaced with the automobile in the United States in the early 1900s, the rate of fatalities per vehicle mile driven was incredibly high. Consider that in 1921, there were over 24 fatalities per one hundred million vehicle miles traveled (VMT). By 2015, almost one full century later, usage of the automobile had increased by over 50-fold, and the fatality rate had dropped from 24 to 1.15 fatalities per one hundred million VMT.

1. Jeff Larson, Surya Mattu, Lauren Kirchner, and Julia Angwin, "How We Analyzed the COMPAS Recidivism Algorithm," ProPublica, May 23, 2016, https://www.propublica.org/article/how-we-analyzed-the-compas-recidivism-algorithm
2. Hill, Kashmir. "Accused of Cheating by an Algorithm, and a Professor She Had Never Met." *New York Times*, May 27, 2022, https://www.nytimes.com/2022/05/27/technology/college-students-cheating-software-honorlock.html
3. James Vincent, "Twitter taught Microsoft's AI chatbot to be a racist asshole in less than a day," The Verge, March 24, 2016, https://www.theverge.com/platform/amp/2016/3/24/11297050/tay-microsoft-chatbot-racist
4. Zhang, Maggie. "Google Photos Tags Two African-Americans As Gorillas Through Facial Recognition Software." Forbes. July 1, 2015. https://www.forbes.com/sites/mzhang/2015/07/01/google-photos-tags-two-african-americans-as-gorillas-through-facial-recognition-software/?sh=2f49e9e713d8
5. Amit Datta, Michael Carl Tschantz, and Anupam Datta, "Automated Experiments on Ad Privacy Settings: A Tale of Opacity, Choice, and Discrimination," arXiv, https://arxiv.org/abs/1408.6491

As automobile miles traveled increased, relative fatality rates plummeted

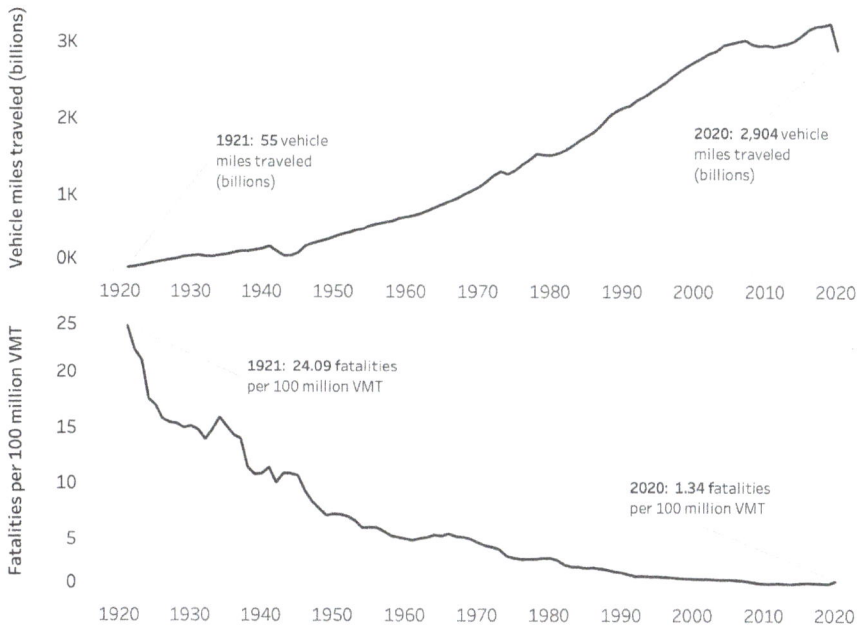

Source: National Highway Traffic Safety Administration (NHTSA)
Data table: Wikipedia - https://en.wikipedia.org/wiki/Motor_vehicle_fatality_rate_in_U.S._by_year

Figure 1.1. Comparison of vehicle miles traveled and vehicle fatality rates in the United States

The reasons for this dramatic drop in fatality rate over that time are obvious. At the beginning, crowded city streets didn't have any signs or traffic lights. By 1930, drivers in just 24 U.S. states needed to get a driver's license, and drivers in just 15 states needed to pass a mandatory driver's exam.[6] In California, it wasn't legally required to wear a seat belt until January 1, 1986.[7] And front airbags weren't mandatory until the 1999 model year.[8]

Take a look at this photograph of a crowded city street in Los Angeles in 1910, and you'll get an idea why the rate of accidents was so much higher than it is today.

6. Elizabeth Nix, "When was the first U.S. driver's license issued?," HISTORY, June 24, 2016, https://www.history.com/news/when-was-the-first-u-s-drivers-license-issued
7. Willie Monroe, "FROM THE ARCHIVE: 1st day of driving when wearing seat belts became CA law in 1986," ABC7 News, January 1, 2021, https://abc7news.com/seat-belts-law-california-wearing/9207393/
8. "Airbags." Insurance Institute for Highway Safety. Accessed May 18, 2023. https://www.iihs.org/topics/airbags

Figure 1.2. Public Domain. Source: University of Southern California. Libraries and California Historical Society (https://doi.org/10.25549/chs-m781)

That's fine for automobiles, but where are we along the adoption curve for data? The digital revolution may have started decades ago with the invention of the transistor in 1947. But when we consider the fact that we Homo sapiens have been on the planet for over 300,000 years, we realize that we're in the early days of adoption of digital forms of data as a means of making decisions and organizing our endeavors and our societies. And with accelerating adoption of data during the early days of the internet in the 21st century, it's quite evident that the accidents are frequent, and they're often pretty brutal.

So we must ask the question: what changes do we need to make in order to bring down the rate of harm done with data? The answer is that we need more robust systems of guardrails and safety mechanisms to ensure that standards of data ethics are routinely considered and consistently met. We need to develop a generation of leaders who are data-savvy enough to know how to prevent the many harmful side effects suffered by innocent people when data is misused.

Sure, some hand-wringing about the current state is in order, but let's cut that part short and get down to business. The data-savvy leader ensures that principles of ethics are not just in place, but also followed, so that data is used to do good rather than to do harm. Let's get more specific. How can you, as a data-savvy leader, make sure sound ethical principles are in place before you rush ahead and put others at risk?

Seek to Use Data for Good

Work with your team to use data for good, to improve the lives of your customers, employees, partners, and other stakeholders.

Very few people (though they do exist) would take pleasure in purposefully and willfully using data to hurt someone. Sure, there are those out there who engage in fraud and theft without giving a second thought to the victims of those criminal activities, but the vast majority of people want to be helpful rather than harmful. Like you and me, they want to wake up every day, and spend their time working for an organization that improves the lives of others in some way. This is likely true of every single one of the employees who work on your team.

Given this almost universal desire to be a force for good in the world, or at least to avoid being a force for evil, there's value in stating the intention to use data to benefit others—customers, shareholders, employees, partners, the general public, and so on. If your team only hears you talk about financial goals or competitive ones, they may assume that the ends justify the means, and that doing something unethical is only bad if they don't get away with it.

As the leader, you need to send a clear message to everyone that data is to be used in constructive ways rather than in destructive ways. A starting point can be to draft and share a motto, credo, or aspirational goal with the team.

Consider the nonprofit organization Viz for Social Good.[9] Their mission statement is: "Our community of volunteers helps mission-driven organizations create social change through data visualization and storytelling." Their vision statement is: "We envision a world where everyone can benefit from data visualization."

---

9. "Viz For Social Good," Viz for Social Good, accessed May 18, 2023, https://www.vizforsocialgood.com/our-mission

Now, doing good with data isn't just a motto for them; it's why they exist. Your team probably has other reasons for existing, but the "do good with data" aspect can be baked right into your team's *raison d'être*, too. To use another metaphor, data ethics can be one of many "pillars" that you put in place to create a structure for your team. Consider the following examples:

- "We use data for good."
- "We use data to help people."
- Like the original Hippocratic oath: "First do no harm…with data." (Latin: *Primum non nocere*)
- Also like the original Hippocratic oath: "First do not hoard (sensitive data)."
- Like the Golden Rule: "We do with others' data as we would have them do with ours."

If you'd like to go beyond a simple motto and develop your own set of principles of data ethics that apply directly to what your team does, the good news is that you don't have to start from scratch. There are a number of helpful sets of guidelines that have been published by reputable organizations that you can use as a starting point.

- The Association for Computing Machinery (ACM) has published their ACM Code of Ethics and Professional Conduct.[10]
- The American Statistical Association (ASA) has published their Ethical Guidelines for Statistical Practice.[11]
- The Manifesto for Data Practices, of which I was fortunate to be a co-author, is currently housed by the Linux Foundation.[12] This document espouses 4 values and 12 principles. Its 4 values are as follows:

---

10. "ACM Code of Ethics and Professional Conduct," Association for Computing Machinery, accessed May 18, 2023, https://ethics.acm.org/code-of-ethics/.

11. "Ethical Guidelines for Statistical Practice," American Statistical Association, accessed May 18, 2023, https://www.amstat.org/your-career/ethical-guidelines-for-statistical-practice

12. Data Practices Authors. "Data Values and Principles." Data Practices Manifesto. Accessed May 27, 2023. https://datapractices.org/manifesto/

1. **Inclusion**: Maximize diversity, connectivity, and accessibility among data projects, collaborators, and outputs.

2. **Experimentation**: Emphasize continuously iterative testing and data analysis.

3. **Accountability**: Behave ethically and transparently, fix mistakes quickly, and hold ourselves and others accountable.

4. **Impact**: Prioritize projects with well-defined goals, and design them to achieve measurable, substantive outcomes.

The 12 principles of the Manifesto for Data Practices are as follows:

"As data teams, we aim to...

1. Use data to improve life for our users, customers, organizations, and communities.

2. Create reproducible and extensible work.

3. Build teams with diverse ideas, backgrounds, and strengths.

4. Prioritize the continuous collection and availability of discussions and metadata.

5. Clearly identify the questions and objectives that drive each project and use {them} to guide both planning and refinement.

6. Be open to changing our methods and conclusions in response to new knowledge.

7. Recognize and mitigate bias in ourselves and in the data we use.

8. Present our work in ways that empower others to make better-informed decisions.

9. Carefully consider the ethical implications of choices we make when using data, and the impacts of our work on individuals and society.

10. Respect and invite fair criticism while promoting the identification and open discussion of errors, risks, and unintended consequences of our work.

11. Protect the privacy and security of individuals represented in our data.

12. Help others to understand the most useful and appropriate applications of data to solve real-world problems."

If you visit the Manifesto for Data Practices website, you can read and sign your name to this online document. At the time of writing, more than 2,200 individuals have done so. It would be great to see this and other initiatives like it grow so that more people become aware of the importance of data ethics.

Being ethical is like climbing a mountain, though. You'll never get to the top unless you really want to get there. It isn't generally the path of least resistance, because being ethical requires us to ask additional questions and to put in place additional measures to prevent harm from being done. Being ethical can involve choosing a path that actually lowers your return on investment in the near term, because you will forgo more profitable routes that would be harmful or unfair in some way. It often involves slowing down, which can be difficult to do in the face of competitive pressures in the marketplace. It's an uphill battle.

As we'll discuss next, merely intending to get there isn't enough. You can shout your fancy data motto until you're blue in the face. But if you don't put it into practice, it'll be a hollow refrain. Worse, it'll give everyone a feeling of doing good, when in actual fact they're doing the opposite.

Consider the Impact of Data-Informed Decisions

Carefully consider the ethical implications of decisions you and your team make when using data, and the impact these decisions may have on individuals and society.

Ultimately, in order to actually apply data ethics, we need to do much more than just say we ascribe to some credo. As nice as it would be just to sign our name on the dotted line and walk away, we won't get off the hook so easily. We have to put the principles of the credo into practice, and we have to behave in a way that's consistent with those principles. To quote the book *Ethics and Data Science*:

> *"Ethics really isn't about agreeing to a set of principles. It's about changing the way you act."* [13]

So how do we act, when it comes to data? Acting a certain way can be either intentional or unintentional. At times we're on autopilot, going through the motions and just doing whatever we're doing without giving it all much thought. Other times we're more deliberate about the actions we're taking, weighing the alternatives, choosing a path to take, and then taking it.

As the leader of a team, you will oversee both types of actions. Your team will make small choices almost automatically, and they will make choices that involve careful consideration and even deliberation. More and more, you will be taking data into consideration as you make these decisions, both small and large, tactical and strategic.

I prefer to use the term "data-informed decisions" rather than the more popular version of the term, "data-driven decisions," because I feel that what should *drive* us is our overall purpose rather than the data itself. Data can be a critical input to our decision-making process, and so can other factors such as intuition, emotions, and ethical considerations. This is particularly true when a person or group of people are actually making the final decision, as opposed to when an algorithm automatically decides.

If data is all that drives your team's decision, what happens if the data points you down a path that would be unfair in some way? Data is often used to score and rank alternatives or to provide quantitative ways of comparing options against each other. But sometimes those scores are derived from data that was collected during periods of unfairness and imbalance.

For example, in October 2018, Jeffrey Dastin of Reuters reported that Seattle-based Amazon decided to scrap an AI recruiting tool that was

---

13. Mike Loukides, Hilary Mason, and DJ Patil, *Ethics and Data Science* (Sebastopol, CA: O'Reilly, 2018).

showing bias against women.[14] A team at Amazon had been creating algorithms to review applicants' resumés. Their goal was to analyze resumés and assign job candidates a score ranging from one to five stars, kind of like the ratings we see on books and products that we purchase on their website. That way, hiring managers could simply sort by the top-rated candidates and hire those individuals, passing over candidates with fewer stars.

By 2015 they noticed, however, that the tool was rating applicants for software development jobs in a gender-biased manner. Since a high percentage of those applying for software developer roles in the previous 10-year period had been men, the model trained itself on their resumés, resulting in poorer scores for resumés that included words more likely to be found on a female applicant's resumé. According to Dastin:

> *"In effect, Amazon's system taught itself that male candidates were preferable. It penalized resumes that included the word 'women's,' as in 'women's chess club captain.' And it downgraded graduates of two all-women's colleges, according to people familiar with the matter."*

As the article mentions, Amazon decided not to use this technology, and they disbanded the team. In other cases, it's not so easy to opt out of data-induced gender bias. Many companies, for example, run Google Ads to promote their offerings. In a 2015 scholarly article, a research team from Carnegie Mellon University and Berkeley University randomly set the browser settings to "female" or "male" and then visited job-related websites, gathering data about what ads were served on those sites.[15]

They found that when the profile was set to "male," ads promoting high-paying jobs were served much more often than when the profile was set to "female." More specifically, an advertisement for the website careerchange.com that read "$200k+ Jobs - Execs Only" appeared 1,816 times to "male" profiles, but only 311 times to "female" profiles, even

14. Dastin, Jeffrey. "Amazon Scraps Secret AI Recruiting Tool That Showed Bias Against Women." Reuters, October 10, 2018. https://www.reuters.com/article/us-amazon-com-jobs-automation-insight/amazon-scraps-secret-ai-recruiting-tool-that-showed-bias-against-women-idUSKCN1MK08G
15. Datta, Amit, Michael Carl Tschantz, and Anupam Datta. "Automated Experiments on AD Privacy Settings." *Proceedings on Privacy Enhancing Technologies* 2015, no. 1 (2015): 92–112. https://doi.org/10.1515/popets-2015-0007.

though both gender settings visited the site a little over 21,000 times in the experiment (p=0.0000053).

It turns out that the website careerchange.com is run by an organization called the Barrett Group. When this research was first published back in 2015, the *Pittsburgh Post-Gazette* reached out to the Barrett Group and had this to say in their July 7, 2015, article "CMU researchers see disparity in targeted online job ads":

> *"Barrett Group president Waffles Pi Natusch said he's not sure how the ads ended up skewing so heavily toward men but noted some of the company's ad preferences might push Google's algorithms in that direction. The company generally seeks out individuals with executive-level experience who are older than 45 years and earn more than $100,000 per year."* [16]

Is it fair that men get served ads for high-paying jobs much more frequently than women? No, it's not, and outcomes like this will only perpetuate an imbalance in our world. Decision-makers at the Barrett Group didn't overtly choose to target males for their ads, and Google didn't deliberately set out to discriminate against women when creating their products. But researchers showed convincingly that this is exactly what happened.

It's not just gender discrimination that can result from data-informed decisions. Other recent studies have raised concerns about data-driven discrimination against minorities in sentencing and recidivism (likelihood of reoffending) of defendants in criminal cases,[17] in setting of home loan interest rates,[18] and more.

The truth is that data is a double-edged sword. It can help us achieve breakthrough results, but it can also help us perpetuate, exacerbate, and effectively "lock-in" unfairness in a wide variety of situations.

16. Deborah M. Todd. 2015. CMU researchers see disparity in targeted online job ads. *Pittsburgh Post–Gazette* (July 2015). http://www.post-gazette.com/business/career-workplace/2015/07/08/Carnegie-Mellon-researchers-see-disparity-in-targeted-online-job-ads/stories/201507080107.
17. Angwin, Julia, Jeff Larson, Lauren Kirchner, and Surya Mattu. "Machine Bias." ProPublica, May 23, 2016. https://www.propublica.org/article/machine-bias-risk-assessments-in-criminal-sentencing.
18. Bartlett, Robert, Adair Morse, Richard Stanton, and Nancy Wallace. "Consumer-Lending Discrimination in the Fintech Era." Journal of Financial Economics, May 29, 2021. https://www.sciencedirect.com/science/article/abs/pii/S0304405X21002403?via%3Dihub.

As your team's leader, it's not enough to have them sign a data ethics manifesto. You have to lead them to think carefully about how your data-informed decisions might affect real people in a negative way. You can lull yourselves into a false sense of security by telling yourselves that you don't intend to cause any harm to anyone. The road to hell, as they say, is paved with good intentions.

Take Data Security and Privacy Seriously

Take data privacy seriously, and work with your team to ensure that sensitive information entrusted to you is secure.

Let's begin this section by remembering that companies are using our own personal data, too. One of the many ways each one of us has been burned by organizations since the advent of the internet is by the way they have mishandled and misused our personal data. The U.S. Department of Labor defines Personal Identifiable Information (PII) as "any representation of information that permits the identity of an individual to whom the information applies to be reasonably inferred by either direct or indirect means."[19] This includes names, addresses, social security numbers, and also combinations of information such as birth date, gender, and race that, when put together, can be used to identify specific individuals.

As a leader, how do you make sure your team is using PII ethically? There are two main topics to consider and get right: data security and data privacy. Data security and data privacy are different but related topics. You can think of them as two sides of the same coin. A simple way to distinguish between them is that data security generally involves *mishandling* of personal data, while data privacy generally involves *misusing* personal data.

First, let's consider **data security**. Organizations possess sensitive PII about each one of us, and these organizations haven't always safeguarded

---

19. "Guidance on the Protection of Personal Identifiable Information." DOL. Accessed May 28, 2023. https://www.dol.gov/general/ppii.

our data well enough, allowing our data to fall into the hands of those to whom we haven't given it. Hackers and other bad actors have gotten access to our data and used that data to commit acts of impersonation, theft, and fraud against us. If an organization doesn't protect our personal data, that amounts to a violation of data security.

Individual data breaches have resulted in millions and millions of people being exposed in ways they don't always understand. In August 2013, a breach at Yahoo resulted in data from over three billion accounts being compromised. In June 2021, data from around 700 million LinkedIn profiles was scraped and aggregated with data from other sites and posted for sale on a dark web forum. This LinkedIn compromise didn't involve a "breach" per se, as no private member data was exposed.[20] But it still amounted to an exposure that users found disturbing and unacceptable.

As a leader, you need to be fully up to speed on all of the measures your organization is taking in order to protect the security of the sensitive data your team is handling. Steps like multi-factor authentication (MFA), use of strong passwords, use of firewalls and virtual private networks (VPN), use of security software and virus scanning, regular updating of software, and receiving cybersecurity training are just a handful of the most basic steps you and your team members need to take in order to prevent the next costly breach.

You don't want your company's name mentioned in the next data breach headline. Reach out to the cybersecurity experts within your organization and make sure you're aware of the latest methods your organization is using to secure your data. Ask yourself what you'd expect a leader at a company that handles your own data to do, and then take it upon yourself to live up to those same expectations.

The second side to this coin is **data privacy**. Organizations possess sensitive PII about each one of us, and they don't always use it in ways that we would approve of or consent to. For example, they may use it to send us communications we don't want to receive in our inboxes. They may use it to target us with advertisements that we don't want to see in

---

20. LinkedIn Corporate Communications, "An Update on Report of Scraped Data." An update from LinkedIn, June 29, 2021. https://news.linkedin.com/2021/june/an-update-from-linkedin.

our browsers. They may give or even sell our data to third parties that we don't want possessing it. They may collect data about us that we haven't even given them permission to collect.

For example, during the COVID-19 pandemic, people around the world turned to video conferencing in order to remain productive in their jobs, as travel and on-site collaboration became restricted. During that time, in April 2020, analysis by the *New York Times* revealed that one of the largest video conferencing platforms in the world, Zoom, was sharing LinkedIn profile details of those who participated in meetings, without those participants being asked for their permission. This was even happening when the meeting participant had logged in anonymously.[21]

The way the platform worked, Zoom users who subscribed to a LinkedIn service called LinkedIn Sales Navigator could enable a feature on Zoom that would allow them to see a LinkedIn icon next to the names of all participants in a meeting. Clicking on that icon would reveal information about the participant such as their name, their employer, their location, and their job title. This would even be possible when the user had logged into the meeting anonymously. At no point were participants made aware that their data was being shared in this way, and this feature of the platform even went around a participant's attempt to withhold such personal information from others in the meeting, such as by logging with the name "Anonymous."

Zoom decided to disable the feature when *New York Times* reporters contacted them to inquire about the way this feature was violating users' privacy, but the situation raises further concerns and doubts about how internet applications collect and share our data. The key concept is **consent**. We should have the opportunity to consent to the specific uses of our personal data, and we should have the ability to withhold that consent if we so choose. If an organization misuses our personal data, that amounts to a violation of our data privacy.

As a leader within an organization, you can stress the importance of data security and data privacy to your team members, and you can ask the hard questions about whether and how consent has been given by

21. Krolik, Aaron, and Natasha Singer. "A Feature on Zoom Secretly Displayed Data From People's LinkedIn Profiles." *New York Times*, April 2, 2020. https://www.nytimes.com/2020/04/02/technology/zoom-linkedin-data.html.

your customers and other partners. You can ask the hard questions about how data is being protected and secured, and you can make sure that sensitive records aren't widely accessible to those on your team who don't need to use them.

Invite and Involve Diverse Voices

Include diverse voices on your team to help you identify and eliminate sources of bias in your data so that people are treated fairly.

When you and your team are using data to make decisions that affect a large number of people, it's important to have a diverse group of team members providing input to the decision-making process. Even when the decision might only affect a small number of people, diversity of input and diversity of voices can help you steer clear of common pitfalls into which we have seen organizations fall over and over in recent years. The fact is that data has been used in many ways that have negatively impacted disadvantaged groups. Let's consider just a few examples.

First, when assessing whether a patient qualifies for a kidney transplant, some transplant centers in the United States, such as Mayo Clinic, directly measure glomerular filtration rate, or GFR, which tells clinicians how well a patient's kidneys are functioning.[22] But this approach is cost-prohibitive or infeasible for other laboratories, so in those cases clinicians have to rely on equations in commercial software to estimate kidney function levels. There are a number of such formulas in use, but they all provide *estimated* glomerular filtration rate, or eGFR, as the output used to guide decisions.

Up until recently, one of the most widely used formulas for eGFR included a race variable with two options, "Black" or "Not Black." Research found that the effect of choosing "Black" in the software was to

---

22. Carlson Kehren, Heather. "Expert Alert: What a Formula Change May Mean for Black Patients in Need of a Kidney Transplant - Mayo Clinic News Network." Mayo Clinic, August 3, 2022. https://newsnetwork.mayoclinic.org/discussion/expert-alert-what-a-formula-change-may-mean-for-black-patients-in-need-of-a-kidney-transplant/.

overestimate kidney function by as much as 16 percent.[23] The result of directly adjusting for race in this way was to block kidney transplants for many Black patients, and to prolong wait times for other Black patients for whom a transplant was approved.

Based on these findings, on June 27, 2022, the Organ Procurement and Transplantation Network (OPTN), a federally mandated network of regional nonprofit organ procurement organizations in the United States, approved the elimination of race-based calculation for determining whether to list a patient as a transplant candidate.[24] Starting on January 5, 2023, kidney programs were required to reevaluate their transplant wait-lists, making corrections to wait times for any Black patients for whom kidney function had been overestimated using a race-based formula.

There are similar commercial algorithms in use in the U.S. healthcare system to guide decisions about conditions other than kidney disease, such as diabetes. These other algorithms don't always explicitly include race as an input variable to determine care, but they can have discriminatory effects anyway, according to research by Obermeyer and colleagues, published in *Science* back in 2019.[25]

When these algorithms are developed using data that reflects inequalities in society, such as poorer health outcomes and lower levels of insurance for individuals in lower income brackets, then disadvantages can be reinforced and they effectively become "systemic." This means such disadvantages apply across the entire system, as opposed to just one part of it.

Another example of how data can perpetuate racial discrimination can be found in a law enforcement approach known as predictive policing. In this approach, police departments attempt to forecast crime rates based on historical arrest data by geographic location. On the surface, this seems to be a brilliant approach: use data to stop crime before it

---

23. Eneanya, N. D., W. Yang, and P .P. Reese. (2019). Reconsidering the Consequences of Using Race to Estimate Kidney Function. JAMA 322(2):113–114. doi:10.1001/jama.2019.5774
24. "OPTN Board Approves Elimination of Race-Based Calculation for Transplant Candidate Listing - OPTN." Organ Procurement and Transplantation Network, June 28, 2022. https://optn.transplant.hrsa.gov/news/optn-board-approves-elimination-of-race-based-calculation-for-transplant-candidate-listing/.
25. Obermeyer, Ziad, Brian Powers, Christine Vogeli, and Sendhill Mullainathan. "Dissecting Racial Bias in an Algorithm Used to Manage the Health of Populations." Science, October 25, 2019. https://www.science.org/doi/10.1126/science.aax2342.

occurs, so our hardest-hit communities become safer. Isn't that exactly what we would want our police to do? Perhaps, but there's a problem.

The key to the problem lies in the fact that a database used to predict future crime rates doesn't include *all crimes* committed in an area. It only includes arrests that were recorded in the database. That seems obvious, but we need to stop and think about why a particular crime might not be included in the database:

- Maybe no police officer was in the neighborhood to observe a crime as it occurred.

- Maybe no one from the community was willing to report it to the police.

- Maybe a crime was observed and/or reported, but the police officer involved decided not to arrest anyone for one reason or another.

There may be other cases in which an arrest was dropped, never recorded, or deleted from the database, either on purpose or by accident. We can think of other possible "gaps between data and reality," as I termed the problem in my book *Avoiding Data Pitfalls*. The point is that every arrest database has "blind spots" such as these, which means that non-blind spots get overemphasized in the predictive model.

In other words, predictive policing can involve a vicious cycle: the more policing happens in a particular neighborhood, the more likely it is that crimes committed in that neighborhood will be observed by the police, and arrests made for those crimes. This results in a higher forecasted crime rate there, and therefore even higher levels of policing, and so on.

If there's any bias at all to the original reason why policing was higher in a particular neighborhood in the first place, then this bias effectively becomes amplified by the algorithm, just like feedback from a microphone and speaker that just gets louder and louder.

Kidney transplants and policing levels are just two of the many ways in which data has been used against minorities in a discriminatory way. We could also spend time considering the way facial recognition technologies have been less accurate for images of darker-skinned women.[26] We could consider cases in which self-driving cars have failed to detect pedestrians who have darker skin.[27] We could even consider the results of an AI-judged beauty contest in 2016 in which all but six of the 44 finalists were White, the others were Asian, and only one had visibly darker skin.[28]

Of course data doesn't only discriminate based on race and skin color. We have already considered unfair treatment based on gender earlier in this chapter, and we could also explore other examples in which age, religion, or any other protected attribute is what's being discriminated against.

The fact is that our data has been collected from a world that includes many imbalances. Our machine learning algorithms reflect the ethical biases in the data used to train them. When we use such algorithms to make decisions, we effectively lock these biases in place, making it harder and harder for disadvantaged groups and individuals within those groups to break free from these biases and achieve fairer outcomes.

So what do you do to address these issues? To start with, you should reach out to your human resources generalist to understand the current level of diversity of your own team, your department, and your organization. Having a diverse team can help you identify these and other types of issues upfront.

If your team lacks diversity today, you can ask others within the organization to review your team's work. It would be ideal if the organization had a "data ethics review board" composed of a diverse set of experts from both inside and outside the organization to provide input and spot ethical issues before they are unleashed on the world. If you don't have such a group, lobby for one to be created.

The authors of *Ethics and Data Science* effectively summarized the overall need for diversity on teams that are using data to make decisions that could negatively impact individuals or groups.

26. Lohr, Steve. "Facial Recognition Is Accurate, If You're a White Guy." *New York Times*, February 9, 2018. https://www.nytimes.com/2018/02/09/technology/facial-recognition-race-artificial-intelligence.html.
27. Samuel, Sigal. "A New Study Finds a Potential Risk with Self-Driving Cars: Failure to Detect Dark-Skinned Pedestrians." *Vox*, March 6, 2019. https://www.vox.com/future-perfect/2019/3/5/18251924/self-driving-car-racial-bias-study-autonomous-vehicle-dark-skin.
28. Pearson, Jordan. "Why an AI-Judged Beauty Contest Picked Nearly All White Winners." VICE, September 5, 2016. https://www.vice.com/en/article/78k7de/why-an-ai-judged-beauty-contest-picked-nearly-all-white-winners.

"Team members should be from the populations that will be impacted. They'll see issues well before anyone else. External peer reviews can help to reveal ethical issues that your team can't see. When you're deeply involved with a project, it can be hard to recognize problems that are obvious to outsiders."[29]

The data-savvy leader understands the importance of inviting and involving diverse voices. By doing so, they can increase the chance that any sources of discrimination in their data will be identified and eliminated upfront. That way, data-informed decisions will be less likely to cause further harm. Such leaders will help us move forward to a more balanced world.

Foster Trust in the Data

Work to ensure that your team is getting access to trustworthy data, and that they trust that the data will help them make better-informed decisions.

As a leader, you want your team to trust the data they're using, but only if the data is trustworthy enough. No data set is perfect, but there's a difference between a data set that has minor flaws and one that has major flaws. Some data sets have flaws that are substantial enough that the data simply can't be trusted—not by you, not by your team, not by anyone. It would be unethical to encourage people to trust something you know to be untrustworthy. It would also be unethical not to evaluate the trustworthiness of something you're telling people to trust.

We can consider the relationship between the data's **trustworthiness** and your team members' **trust** in the data using the two-by-two grid shown in Figure 1.3. The horizontal axis encodes the level of trustworthiness of your team's data, and the vertical axis encodes the team's level of trust.

---

29. *Ethics and Data Science*, p. 36

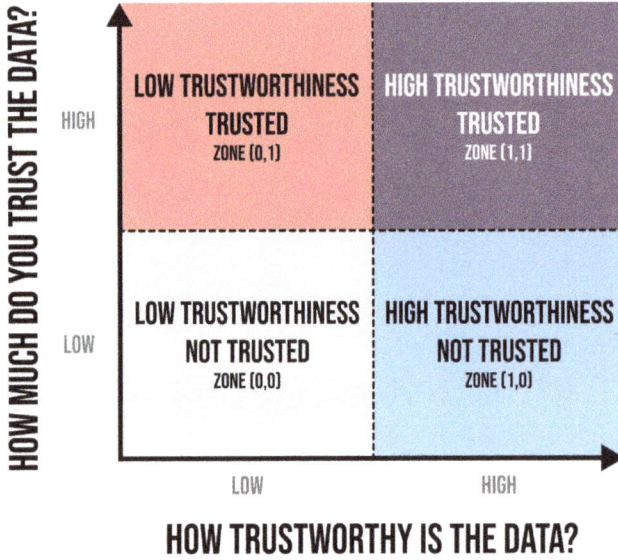

Figure 1.3. Four Zones of Data Trust

This diagram presents us with two independent, binary options, forming the Four Zones of Data Trust. Of course we can acknowledge and appreciate that, in reality, neither option is binary, and we can find many gradations on both axes. After all, data sets can be trustworthy in varying degrees, and the team's trust can be characterized by varying shades of gray as well. But humor me for a moment, and assume that you can place your team's current situation in one of these four zones.

If your team is in Zone (0,0), then they're right where they should be. Their data is woefully problematic, and they know it. If your team's data has major issues with it, you don't want to tell them that they should trust it anyway. That would be like telling a passenger in your car not to worry even though you're driving blindfolded. You'll lose all credibility, and you'll actually be putting them in harm's way.

The best approach in Zone (0,0) is to acknowledge that the data isn't going to be useful in its current state, and then work with your team to create a list of the problems you have with it, along with a "data wish list." We'll consider more of the specific ways a data set can be problematic in the third chapter. After you have your lists, you need to work with your partners in IT and across the organization to advocate for improvements to your team's data. It's your job to fight for your team in this way, and you need to find a productive way to do it. That's what a data-savvy leader would do.

If your team is in Zone (0,1), then unfortunately they're in for a rude awakening. They've been cruising along thinking everything is hunky dory, but the truth is that it's not. They're like a patient in the Middle Ages who's undergoing bloodletting: they trust that the treatment is helping them, but in the vast majority of cases it's actually making the situation far, far worse. Flawed data can do a lot of harm.

How did they end up in such a deluded state? It's easy and quite common: they heard some marketing pitch that data is the wave of the future, and they bought into it lock, stock, and barrel. They didn't stop to question the hype. Maybe they went to an exciting conference organized by a software company, took home some data buttons and data T-shirts, and now they're quite high on data. The problem is, the enthusiastic speakers and sales reps at the conference failed to mention that the software only helps if the data itself is in good enough shape to actually use.

As a leader with a team in Zone (0,1), you have the unenviable job of bursting their bubble of bliss relative to their data. Watch out: they could come crashing down and feel utterly disillusioned about data. You should use the same tools you'd use for teams in Zone (0,0): create a list of problems and a "data wish list." Once they see the blemishes that are really there, you can help them come up with a game plan to adjust their processes while you work to get an improvement timeline from the people at your company who own and maintain the data your team has been using.

Leaders of teams in Zone (1,0) have the opposite problem as those of teams in Zone (0,1). Instead of naively believing that bad data is good, their team members cynically feel that good data must be bad. Why do they have such a negative opinion of the data? Here are a few reasons that the data itself can get a bad rap:

* Maybe one specific source of data they've used is a complete disaster, and now they suspect that all other data sources must be trash, too.

- Maybe the data you want them to use was once unacceptable, and they're simply not aware of the recent improvements that have been made to it.

- Maybe someone who presented the data to them in the past made a glaring mistake in their analysis, and now they think the data itself is bogus.

- Maybe leaders at your organization have used data in excessively political ways recently, bending it a little too much in order to suit their agenda.

The bottom line is that data has a reputation within your organization, and that reputation might not be so great. Sometimes data deserves its bad reputation, as in Zone (0,0) and Zone (0,1). Other times it has been judged unfairly, or deserves another chance, as in Zone (1,0). When this is the case, your job as a leader is to get your team members to be willing to step out onto a platform they think is going to collapse underneath them. How do you do that?

There are two sides to this coin: an intellectual side and an emotional side. Opening up a dialogue with your team members can help you address both sides of the coin. Good, old-fashioned listening skills come in handy here. Listen to your team members relate what they think is wrong with the data, and listen to why they have reservations about making use of it. If their lack of trust is mostly due to emotional reasons like fear, then seeking to convince them using rational arguments won't help.

No matter which of the three problematic zones you're in, what you're hoping to do is move to the top right quadrant—Zone (1,1). This is the "Zone of Appropriate Trust." Team members know that the data, while not perfect, is good enough to help them get the job done, so they use it to do just that.

Of course these four quadrants, or zones, constitute an overly simplistic model. If you feel there might be incremental value in your case, you can create a three-by-three grid with "low," "medium," and "high" trust and trustworthiness levels. My guess is that going beyond this level of gradation will invoke the law of diminishing returns. Don't overthink it.

No matter how granular you choose to make it, remember that in order to foster appropriate trust in the data, you need to consider the quality of the data itself, what your team members think about it, and how they feel about it.

Encourage Raising the Red Flag

Ensure that your team members feel confident that, if they discovered that your team was using data in an unethical way, they could raise a warning flag without fear of retribution.

Unfortunately, it's not so easy to ensure that individuals won't ever be harmed by data that your team collects about them. One reason for this is that data can be used in ways that are very different from what was originally intended. Your team may have the most ethical of intentions for collecting sensitive data, and you may go to great lengths to inform those affected about the intended uses of the data, and gain their consent. But once that data has been collected and stored somewhere, you don't know who else will gain access to it down the road, or what they will do with it.

For example, it's easy to understand why a hospital would have good reasons to collect data about a patient's medical care. That's a legitimate use case. And it makes sense that an app that a person uses on their mobile phone to track their health in some way would need to store that data for them so they can use it. But what happens when a person's health data is obtained by those whose reasons for wanting it aren't in alignment with the patient's own interests? Every year, health data is stolen, given away, sold, and combined with other data in ways that the patients who are providing it would find disturbing.

Data sets that contain attributes that can be used to uniquely identify someone, such as their name and Social Security Number, are often protected quite carefully. There is, however, an entire marketplace for data sets that have been anonymized, or "de-identified," meaning all Personal Identifiable Information (PII) has been removed. These anonymized data sets pose no privacy threat to individuals, right? Wrong. Even supposedly anonymous data can be traced back to specific individuals. How is this possible?

Data sets contain various attributes about individuals, such as their gender, birthday, and marital status. It isn't abnormal for data sets to contain hundreds or even thousands of such attributes, including eye color, car color, and favorite color, among many others, most of which have nothing to do with color. Researchers at the Université catholique de Louvain in Belgium and at Imperial College London in the UK found that 99.98% of Americans can be uniquely identified using just 15 attributes.[30]

These researchers have freely shared the code they used to develop their statistical model, and they've published an online tool that you can use to find out how likely it would be for someone to identify you using even fewer attributes than that. When I input my own gender, birthday, marital status, and zip code, their model estimates that someone would have an 86% chance of identifying me using those four attributes alone. Add to that the number of vehicles I own, and the probability shoots up to 96%.

It's even possible for a single attribute to be used to uniquely identify someone in a supposedly anonymous data set. At the 2017 DEF CON Conference in Las Vegas, journalist Svea Eckert and data scientist Andreas Dewes teamed up to deliver a presentation titled "Dark Data,"[31] in which they shared the result of their analysis of a data set they were able to access that contains the browsing history of a sample of three million Germans. Even though the data set was supposedly anonymized, some URLs in the data set contain the username of the person who is visiting them.

One example is Twitter Analytics, whose URL currently takes the form: https://analytics.twitter.com/user/USERNAME/home, where "USERNAME" is the Twitter handle of the user, such as "DataRemixed" for my own account. You can only get to that page using a browser that's currently signed into your Twitter account, so this URL can effectively

---

30. Luc Rocher, Julien Hendrickx, and Yves-Alexandre de Montjoye. "Estimating the success of re-identifications in incomplete datasets using generative models." Nature Communications 10 (2019) 3069.
31. "Def Con 25 - Svea Eckert, Andreas Dewes - Dark Data." YouTube, October 20, 2017. https://www.youtube.com/watch?v=1nvYGi7-Lxo.

be used to identify a specific person in the data set. Find a URL like that, and you now know the person's entire browsing history by looking at other websites visited by the same individual.

What was the harm done by this, as well as other deanonymization techniques Eckert and Dewes used? Well, they were able to identify an elected politician's prescription medication, the IP address of a suspect in a criminal investigation, and a judge's pornographic website views. These individuals all gave consent for their website history to be collected by one of many free browser extensions. They likely clicked through some lengthy terms of service agreement like each of us do multiple times per month. If they had bothered to look closely at the verbiage in the agreement, which I'm willing to bet they hadn't, they might have been reassured by a clause stating that the data collected would be entirely anonymous. Not so much.

So what's a leader to do in order to safeguard people from such unintended uses of the data their team collects about them? There's no perfect solution. Data collected is data at risk, so, for starters, "do no harm" becomes "do not hoard." Only collect and store personal data for which the person has given you consent, and for which you have a legitimate use that's in alignment with their own interests. Not sure if a particular attribute is useful in any way? Don't collect it.

This advice flies in the face of the conventional wisdom to collect everything and figure out how to use it later. The reality is that there's a very real cost to both the individual and to your organization for holding on to every scrap of personal data. This cost can be hidden, and it might only surface much later, say, after a breach. So only collect personal data if there's a legitimate business use for doing so.

One other step you can take as a data-savvy leader is that you can make sure that everyone on your team knows that there will be no retribution for anyone who "raises the red flag" about a potential issue that they see. If you're a middle manager, you need to get assurances from those above you in the organizational chart that your team members will be protected if they come forward with a concern about data ethics. Your team members need to be empowered to notify you about problems, and when they do raise the flag, there can be no backlash against them. And when they raise the flag, you need to bring everything to a halt until the situation is assessed and resolved.

In their book *Ethics and Data Science*, co-authors Mike Loukides, Hilary Mason, and DJ Patil draw inspiration from the "andon cord" technique implemented to great effect by Toyota in the now famous Toyota Production System. The andon cord was a physical cord that any person on an automobile production line could pull if they noticed a defect, a supply outage, or any issue that could result in poor quality. Pulling this cord would stop the production line entirely, and people would gather around to understand the issue and work together to come up with a resolution. Loukides, Mason, and Patil make the following connection to the world of data ethics:

> *"Any member of a data team should be able to pull a virtual 'andon cord,' stopping production, whenever they see an issue. The product or feature stays offline until the team has a resolution. This way, an iterative process can be developed that avoids glossing over issues."*[32]

As someone who implemented aspects of the Toyota Production System earlier in my career as a practitioner of Lean Sigma, I can say that this could be an effective approach. At first, and by definition, the andon cord slows things down. No one wants to stop anything when there is so much pressure to go, go, go. But you should think of this approach as an investment.

After a few such stoppages, the quality gradually begins improving, and the need for additional stoppages starts to go down. In the world of data privacy, I'd argue that what's needed is not just a mechanism to *stop* a product or process, but also to pull something *out of production* that someone on your team discovers or suspects to be harmful. In many cases, the costs of getting it right upfront could be far less than the costs of dealing with a data ethics fiasco later.

Gain alignment from executive leadership that employees who raise concerns about data ethics won't be punished. Create a process whereby they can raise the flag, or "pull the andon cord." Maybe you implement an email alias, or an online form, or a community workspace, or an

---

32. *Ethics and Data Science*, p. 34

instant messaging channel. Then work with your executive leadership to form a Data Ethics Review Board to consider all concerns raised on this channel. Populate this board with a diverse set of experts who will be less subjected to the pressures that the organization is facing, such as experts who do not work for the organization.

You may still do harm with data. But this way you'll be far more likely to become aware of it, and you'll be in a far better position to fix it and prevent it from happening again.

Discover and Fix Data Errors

When your team discovers that errors have been made with data, fix the errors and deal with the consequences, rather than hiding or ignoring them.

Mishandling and misusing private records aren't the only ways you and your team can do harm with data. The unfortunate but unavoidable truth is that data is often very dirty, and using it "as is" can result in problems galore. In 2016, IBM estimated that poor quality data cost U.S. businesses $3.1 trillion.[33] In 2021, Gartner shared that poor quality data costs organizations $12.9 million on average.[34]

I can't vouch for those estimates, and I do wonder whether it's even possible to estimate the hidden costs associated with bad data that's never detected. After all, we don't know what we don't know. But based on these estimates, and on my own experience, I feel comfortable saying that data errors can be very costly indeed.

Data is like a kitchen counter. Sometimes you walk into the kitchen and you see a huge mess on the counter—maybe a cup of coffee has spilled all over it. You can't miss it, and even the laziest college student would have a difficult time using the counter in that condition. There's no choice but to clean it up.

33. Redman, Thomas C. "Bad Data Costs the U.S. $3 Trillion per Year." Harvard Business Review, September 22, 2016. https://hbr.org/2016/09/bad-data-costs-the-u-s-3-trillion-per-year.
34. Sakpal, Manasi. "How to Improve Your Data Quality." Gartner, July 14, 2021. https://www.gartner.com/smarterwithgartner/how-to-improve-your-data-quality.

Other times, the kitchen counter looks great at first blush, but a closer look would reveal salmonella bacteria left over from the preparation of last night's chicken cacciatore. You place your hand on what you think is a perfectly clean counter and later that day you find yourself in very bad shape. Data can be like that too, unfortunately.

In *Read, Write, Think Data*, I wrote about three different kinds of dirty data: inaccurate data, inconsistent data, and duplicate data. In brief, **inaccurate data** includes readings that deviate from the "true" value. Measurement devices commonly have a tolerance rating that tells you how much greater or less the reading can be from the true value. Your team's data sets can also contain inaccurate values due to rounding errors, missing decimal points, mismatching units of measure, faulty calculations, and a myriad of other root causes. These kinds of errors can be very costly.

Inconsistent data can result whenever the process to generate the data doesn't produce the same or similar values every time. In quality control, measurement systems might not be very **repeatable**, meaning the same person doesn't get identical results when measuring something twice. And they may not be very **reproducible**, meaning two people get different results when they each take a turn measuring the same object. Additionally, inconsistencies can result in data entry when people enter different values that mean the same thing. One extreme example is the 57 ways people entered text to mean "Philadelphia" in the 2020 U.S. Paycheck Protection Program application form. Manual data entry is replete with typos and inconsistencies like this.

PHIADELPHIA	PHILADELPHAI	PHILADLPHIA
PHIALDELPHIA	PHILADELPHI	PHILADPHIA
PHIDELPHIA	PHILADELPHIA	PHILADRLPHIA
PHIELADELPHIA	PHILADELPHIA PA	PHILAELPHIA
PHIILADELPHIA	PHILADELPHIA,	PHILDADELPHIA
PHILA	PHILADELPHIA, PA	PHILDADLPHIA
PHILA,	PHILADELPHIA'	PHILDAELPHIA
PHILAD	PHILADELPHIAP	PHILDELPHIA
PHILADALPHIA	PHILADELPHIAPHIA	PHILDEPPHIA
PHILADEDLPHIA	PHILADELPHILA	PHILIADELPHIA
PHILADELAPHIA	PHILADELPHIOA	PHILIDELPHIA
PHILADELHIA	PHILIDELPIA	PHILLA
PHILADELHPIA	PHILADELPOHIA	PHILLADELPHIA
PHILADELLPHIA	PHILADELPPHIA	PHILLY
PHILADELOHIA	PHILADEPHA	PHILOADELPHIA
PHILADELPH	PHILADEPHIA	PHLADELPHIA
PHILADELPHA	PHILADEPHILA	PHOLADELPHIA
	PHILADEPLHIA	PHPILADELPHIA
	PHILADERLPHIA	PIHLADELPHIA
	PHILADLELPHIA	
	PHILADLEPHIA	

Figure 14. Different spellings of "Philadelphia" in 2020 PPP applications

A third way data can be dirty is that it can contain **duplicate records**. Duplicate data can result from poorly written SQL queries, from copy-and-paste mistakes in spreadsheets, from repeated scraping of the same web pages, and simply from identical entries into a data entry form. It can also sneak into your data sets, disguised as "total" or "grand total" rows that sum up (and effectively repeat) multiple values in a table. Using data that contains duplicate records can result in double counting, over-charging customers, underordering supplies, faulty rankings, and more. The list of symptoms that can result from this same underlying malady goes on and on.

As a data-savvy leader, you need to embrace the fact that your data is dirty. The question isn't whether or not your data is dirty, but rather how dirty is it? Once you and your team assess the quality level of your data, you can determine whether it's good enough to answer certain questions, or whether you need to clean it up first. It's unethical to proceed and use the data anyway even though you know it has egregious errors. And it's unethical to stick your head in the sand and ignore the possibility that data errors might exist.

What happens when you discover you've gotten it wrong? Let's say a data error is discovered after you've already used it to make a decision, or after you've already communicated the data to a stakeholder. What does a data-savvy leader do in that situation? They correct it, and then they go back and figure out how to prevent another such data error from falling through the cracks again in the future.

In September 2022, Business Insider reported that Amazon made an error in the calculation of some employees' compensation, and they needed to communicate the bad news to these employees that they'd be receiving less than they were originally told.[35] The alleged reason for the error was that they had used an outdated version of the company stock price in the formula. The stock price had declined over the course of the year, and a lower, more recent stock price evidently should've been used in the compensation calculation. They corrected the data error

---

35. Long, Katherine. "Leaked Email Reveals That Amazon Is Walking Back Employees' Raises after an Internal Bug Miscalculated Their Compensation." Business Insider, September 22, 2022. https://www.businessinsider.com/amazon-promotion-raises-compensation-bug-2022-9.

even though it meant many managers and employees had to have some uncomfortable discussions.

A hypothetical question to consider is whether the company would have corrected the error if the opposite mistake had happened, and a higher stock price should have been used in the compensation calculation. It's one thing to correct an error when such an act benefits the party who made the mistake in the first place. But are you willing to issue a correction notification when the error is not in your favor?

If data ethics is truly important to you, you'll make the correction whether it benefits you or not. To complicate things somewhat, an argument can be made that if you make a mistake with data that benefits someone else at your own expense, there are cases when making the correction would not be the right thing to do. Perhaps you undercharged for a product or service that has already been delivered. In most cases, charging the remaining amount when the buyer made their decision on the miscalculated price likely wouldn't be appropriate.

Ethics Summary

Hopefully I have made a convincing argument that data ethics is a critical area of consideration that you as a leader can affect with your words and with your actions. Additionally, I've sought to make the case that you need to work on this important topic as early in the journey as possible. Putting the cart before the horse is not a good idea.

In summary, here are seven questions for you to consider when you evaluate your own leadership relative to data ethics:

1. How are you stressing the importance of using data for good, and not for harm?

2. How are you considering the impact of the decisions you're making with data?

3. What safeguards have you put in place to protect data security and data privacy?

4. Are you inviting and involving diverse voices to the data dialogue?

5. What are you doing to foster warranted trust in the data?

6. How do you encourage team members to raise the flag when they spot an issue?

7. When you become aware of data errors, do you assess and fix as necessary?

Chapter 2

PURPOSE

"Efforts and courage are not enough without purpose and direction."

–John F. Kennedy

Resolving to use data for good rather than for harm is a necessary first step in becoming a data-savvy leader. It won't be easy for you to put that resolution into practice, though, and you will have to keep working at it. As you continue down the ethical road, you'll also want to think carefully about whether you're using data to move toward achieving your team's purpose.

This assumes, of course, that you know what your team's purpose is. Do you? Why does your team exist? Perhaps there is more than one reason. Once you know your team's purpose, you can find out whether you have the data you need. Furthermore, you can use this awareness to find ways to put your data to good use. As you'll see in this chapter, you can also use your purpose statements to derive goals and objectives that hit home for each of your team members.

Over the course of my career, I have heard many talented data professionals complain that they were hired to do great things with data, but they ended up working on "pet projects," or activities that made no difference at all. This is a surefire way to demotivate someone who has highly sought-after data skills. They will leave your team and find one that lets them put those skills to better use.

You want to avoid such meaningless "boondoggles" and instead focus on ways you can empower your team to use data to actually "move the needle." In this chapter, I'll provide seven recommendations aimed at helping you align your data to your purpose. I'll leave you with seven associated questions to ask yourself so that you can assess your own data literacy acumen as a leader.

Shine the Spotlight on Your Team's Purpose

I make sure that my team members are intimately familiar with our purpose, and who benefits from our efforts.

Considering the title of this book, it's highly likely that you purchased it and are reading it now because you want to be a better leader. Maybe you even want to be a great one. That's a worthwhile goal, and I commend you for that.

Depending on your mindset, it can also be an overly self-centered focus, however. It's possible and perhaps even likely that a lot of one's ego is embedded within their motivation to become a great leader. The same can be said for the reason I'm writing this book, or the reason we apply ourselves to any task for that matter. Ego and reputation tend to be a part of the equation, whether we are willing to admit it or not.

We've all worked for that boss who is obsessed with their own leadership skills, or, more precisely, with their own reputation as a leader. We tend to have an innate ability to sniff out the person who is more interested in their own well-being and progress than anyone else's.

The human ego is a fantastic source of energy, but it's also unruly. In that sense, it's like a wild stallion. In order for you to become a truly great leader, you need to tame and harness your ego, and shift the focus off of yourself. What do you focus on instead? You focus on the *who*, the *where* and the *why*:

- *Who* are the individuals on your team?
- *Where* are you going together?

- *Why* does that journey and destination matter?

In a word, great leaders focus on *purpose*. A team exists for some purpose. So what's your team's purpose? How are you getting your team members to think about that purpose, and about their role in fulfilling it? As the leader, you play an important role in fulfilling your team's purpose, too. Your role may require you to be at center stage. Your team may need you to be a mouthpiece for a movement, or a standard-bearer for a cause, for example. Even in such situations, though, it's possible to keep the focus on the team's purpose, rather than on yourself.

You can talk about purpose all day, and say how much it matters to you. Ultimately, though, people judge what matters to you by what you do. And they judge your true motivations by who gets the credit when your team wins. I love this passage from Stephen Mitchell's translation of the *Tao Te Ching*, a classic Chinese text written around 400 BCE and attributed to a sage named Lao-Tzu:

> *"The Master doesn't talk, he acts.*
> *When his work is done,*
> *the people say, "Amazing:*
> *we did it, all by ourselves!"*[36]

It's easy to get confused about the meaning of the various terms we use to describe our aspirations, and the things we're trying to do or achieve: purpose, mission, vision, goals, objectives. The reason for this is that these words are all related to each other, and there's some overlap in how they're used.

Let me attempt to articulate my own understanding of these terms based on how I've seen them best used in an organizational context. I'll provide some examples from my own company, Data Literacy, to make the meanings clearer. The list of terms starts at the highest and most general level and works its way down to the lowest and most detailed level. I'll call these "The 5 Levels of Team Aspiration."

---

36. Lao Tzu, *Tao Te Ching*, Ch 17, Stephen Mitchell (Translator) https://www.amazon.com/Tao-Te-Ching-Perennial-Classics/dp/0061142662/

- **Vision**: An idea you have about a future state of the world that you want to bring about; you can think of vision as where you want to end up. With vision, the focus is on the future, and on how you want that to look.

  - *At Data Literacy, we envision a world in which everyone shares a common language of data, where people of all backgrounds and walks of life can use data to work together.*

- **Purpose**: The reason your team exists; you can think of your purpose as the "why" behind your actions, and it only applies if your vision is not yet a reality. With purpose, you put the focus on the core need you're trying to meet or resolve. When your vision is reality, you have fulfilled your purpose.

  - *Data Literacy exists because more and more people need to use data to solve their problems, but too many people simply aren't comfortable using data yet. Our purpose is to help close this greatest educational gap of our time.*

- **Mission**: An ambition or a calling that drives your activities; you can think of a mission as the approach you're taking to make your vision a reality. With mission, the focus is on today, and on how you're making a difference right now.

  - *At Data Literacy, our mission is to teach people all over the world how to speak the language of data.*

- **Goal**: An outcome you'd like to achieve that's broad or high level; you can think of goals as your targets for the year. With goals, the focus is on this year, and on how you'll measure success.

  - *At Data Literacy, our goal this year is to reach over 20,000 people around the world with our new courses and our free educational content.*

- **Objective**: A more specific action that you'd like to carry out in order to achieve your higher-level goals; you can think of objectives as the things you need to do in order to attain your goals. With objectives, you focus on what meaningful tasks and projects you are setting out to complete.

○ *This year, we will create and launch four brand-new courses, publish 50 blog posts, and record, edit, and post 50 instructional videos.*

As your team's leader, you need to put the focus on all five of these levels of aspiration, starting with your team's purpose, your reason for existing. In reality teams have multiple reasons for existing. It's important to acknowledge that fact, and to bring a balanced view to the table when you discuss your team's purpose with them.

Lead a conversation with your team about their purpose relative to four different areas that we'll call "The Four 'Relatives' of Team Purpose."

- **Relative to the Mission**: Why does your organization need your team in order to achieve its mission? The purpose of a supply chain team is different from the purpose of a customer service team. Articulate why your team exists in terms that connect to the broader organization's mission.

- **Relative to Customers**: Why do your customers need your team in order to get their own needs met? Again, the way a supply chain team meets the needs of the end customer is different from the way a customer service team meets their needs. If you lead a team within a nonprofit organization or a government agency, substitute the word "customer" with the word "beneficiaries," "taxpayers," "citizens," or whatever word you use to refer to the people you directly serve.

- **Relative to Shareholders**: Why do the owners, investors, and shareholders need your team in order to realize a return on their investment (ROI)? Keep in mind that you won't always measure or express this ROI in financial terms. People who give grants to foundations are looking for a particular return on their investment too, as are voters who put an elected official in office. What purpose does your team fulfill in bringing these hopes to fruition?

- **Relative to the Employees**: Why do your team members themselves need to be on the team in order to meet their own needs? We all want to work for organizations that inspire us and motivate us, whose purpose, mission, and vision are all meaningful to us. But we have our own immediate needs too: we need to feed and clothe our families. And we need to grow our careers.

Here is the key point that I'd like to make as we wrap up this section: If you don't continually shine the spotlight on your team's purpose, it doesn't matter how technically advanced you are or how well you're able to wrangle or visualize data. Without purpose, you may be engaging your team members' minds, but you won't be tapping into their hearts. And you won't go nearly as far as a purpose-centered leader will go.

Measure What Matters in Helpful Ways

Directly measure how well your team is doing what you say you want to do, and achieving what you say you want to achieve.

The data-savvy leader uses data to track and measure how well their team is doing relative to their goals. That's conventional wisdom, at this point. I imagine that very few people would challenge that statement. Nowadays almost everyone works in an environment with key performance indicators (KPIs), goals and objectives (G&O), and regular team meetings in which their boss tells them how well they're doing against the annual operating plan (AOP). You know it's a common experience when the acronyms have become common vernacular.

So I won't spend this section telling you the many and various ways you can measure and track anything. Yes, you can track on-time completion rates, percent-to-plan output, quality levels, accuracy levels, and satisfaction scores galore.

The question I'd like to address isn't how to measure progress, but rather how to do it well. This second question implies an appreciation that there are ways to do it poorly. To consider the different ways to measure progress, and to present my thoughts about what to do and what not to do, I'll use this diagram that separates performance metrics into three categories: quantity, quality, and perception.

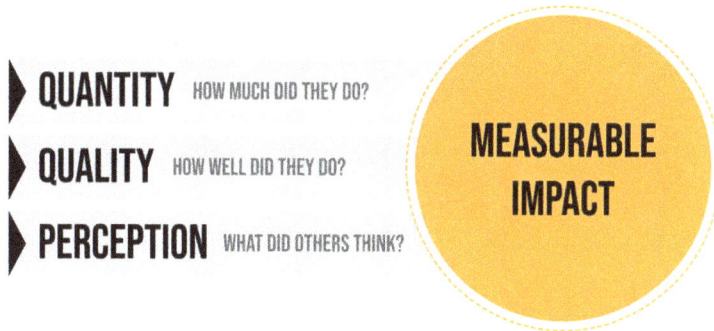

QUANTITY HOW MUCH DID THEY DO?

QUALITY HOW WELL DID THEY DO?

PERCEPTION WHAT DID OTHERS THINK?

MEASURABLE IMPACT

Figure 2.1. The Three Ways of Measuring Performance

Measuring Quantity

The first and most basic way we can think of measuring progress is to count how many times a person or a group has done some activity in some span of time.

- How many leads did you convert into sales?
- How many shipments did you deliver to customers?
- How many projects did you complete on time?
- How many support tickets did you close?
- How many blog posts did you publish?

For roles that involve a high percentage of working time devoted to carrying out recurring processes, counting or measuring activities or output can be a very important way to gauge performance. As a leader, you may oversee a team that's responsible for a process that's critical to the operations of the business. That makes you the **process owner**, so you need to know if your process is running smoothly, and you need to know if your team is getting the job done. This is very basic, and such tracking predates the digital revolution.

Maybe you lead an accounts payable team that processes invoices. How many invoices have they processed this week, this month, or this year? Is there a growing backlog of invoices to process? If so, you may need to address that process to increase throughput, or you may need to hire new employees to handle the higher workload. If you're really on top of your quantity metrics, you can move from a reactive mode to a proactive one by forecasting the growth of invoices and getting in front of the problem before it happens.

Maybe your team recruits new employees, or onboards those new employees. Each employee is different, but your team performs many of the same steps each time. Maybe your team runs conferences and events. Each event has its own unique elements and challenges, but, once again, your team follows very similar steps to plan and run each one. You can count many aspects of your team's processes to find out whether your team is running a well-oiled machine.

Sometimes you're counting activities that you can control, such as the number of emails that you sent to a distribution list. Other times you're counting outcomes that you might be able to affect, but you don't directly control, such as the number of people who clicked on a link in those emails you sent. These quantity metrics are useful and even necessary if you want to be aware of how your team is doing relative to any goals that can be expressed as quantities or measurements of an *amount*.

There are a few ways in which quantity metrics can go awry, though. The first involves highly creative roles. Counting everything that a person in a highly creative role does in the act of creating can be very demotivating for that person. Writers, artists, designers, architects, musicians, and many other creative professionals can go through periods of stagnation before arriving at a eureka moment, after which an incredible amount of value flows from their brains. Badgering them with messages that their output metrics are down might not be helpful.

Shortly after Elon Musk took over Twitter in 2022, he laid off half of the company's global workforce of 7,500. During this period, software programmers were told to physically print out their lines of code and bring them to employee review meetings. The implication seemed to be that the programmers who wrote the fewest lines of code had been the least productive and would be let go.

It isn't entirely clear if Musk and his staff were actually using such a "**force ranking**" system or not, but the insinuation itself triggered a huge backlash in the programmer community. Many expressed their opinion that counting lines of code isn't a valid way to measure how much a programmer has contributed to a team.[37] It's possible that a programmer has been engaging in important planning, delegating, or troubleshooting tasks. They may even have been responsible for what's called "negative code," which involves *removing* lines of code that include bugs or that are involved in features that are being removed from production.

The important point to consider is that coding, like many activities, is a creative endeavor, and quantity metrics often aren't the most important ones. They're certainly not the only metric you should consider when measuring their contribution. According to the popular quote often attributed to Microsoft founder Bill Gates, "Measuring programming progress by lines of code is like measuring aircraft building progress by weight." The metric may tell you how big the product has become, but it doesn't tell you how well it will fly.

It's also important that you only measure activities that are critical to your team's success. If an activity doesn't make a difference, then don't shine the spotlight on it. Speaking of Microsoft, in 2019, when the company first released the "Productivity Score" feature within their Office 365 suite of products, they gave managers the ability to track each individual employee's activities. Who had sent the fewest emails? Who had replied in group chats the least? Who had contributed the least to shared documents? Built-in quantity metrics—what could be better?

There was a problem, though. Privacy advocates rightly expressed concerns about what many called "workplace surveillance."[38] So, in 2020, Microsoft removed the ability to see individual user names in the Productivity Score feature, leaving only aggregated usage data about the overall organization.[39] Perhaps if they had followed a more rigorous data

37. Garner, Bennett. "Elon Musk Just Force-Ranked Twitter Engineers & Fired the Bottom." Medium, November 5, 2022. https://medium.com/developer-purpose/elon-musk-just-force-ranked-twitter-engineers-fired-the-bottom-507ab35b659a.

38. Hern, Alex. "Microsoft Productivity Score Feature Criticised as Workplace Surveillance." *The Guardian*, November 26, 2020. https://www.theguardian.com/technology/2020/nov/26/microsoft-productivity-score-feature-criticised-workplace-surveillance.

39. Spataro, Jared. "Our Commitment to Privacy in Microsoft Productivity Score." Microsoft 365 Blog, December 1, 2020. https://www.microsoft.com/en-us/microsoft-365/blog/2020/12/01/our-commitment-to-privacy-in-microsoft-productivity-score/.

ethics review process as we discussed in the previous chapter, they would have avoided the fiasco altogether. But I do feel that Microsoft made the right decision to change the feature in this way.

I see another problem, though. Even beyond the privacy concern, I wonder whether a leader should really want their team members focusing on *how many times* they're doing these office-based activities. If I put myself in the shoes of one of the team members, I imagine such metrics could prompt me to perform a bunch of unnecessary activities to make my metrics look good. In other words, it sounds to me like a recipe for busy work, and doing for the sake of doing.

But your customers aren't going to be willing to pay more for a product if your team writes more emails while developing it. Can you imagine how they'd respond to an announcement that you'll be raising prices by 20% for the next generation of a product because your team wrote 20% more emails than when they developed the previous model?

Alas, such a rationale would fall flat. After all, you won't take market share away from your competitors just because you spend more time in meetings than they do. That would be like telling a basketball team that the winner won't be determined by the final score of the game, but rather by which team has the fastest average running speed on the court.

If you measure something, your team members will do it. They might be tempted to do it a lot. So only measure activities that matter. Only count what really counts. You get what you measure, so you'll need to balance quantity metrics with another set of important metrics: quality metrics.

Measuring Quality

Quality metrics tell us whether or not something is good enough. Did it meet a set of specifications or requirements? The answer to this question can be expressed in a variety of ways. For any single unit of output, the answer can be expressed as a binary: yes or no, pass or fail, good or bad, acceptable or unacceptable, "ship it" or "scrap it."

When we apply this binary approach to a batch or a group of units produced over a specified period of time, we can obtain a **yield**, or a pass rate. To calculate yield, we count the number of "good" units and divide by the total number of units ("good" plus "bad") made in the same period of time. Multiplying that quotient by 100 turns the proportion into a percentage. There are a few special types of yield to consider.

A **first-pass yield**, or FPY (also known as a throughput yield, or TPY) tells us the percent of all units produced in a process that meet specifications without requiring any rework. **Rework** is necessary whenever something being made or done needs to be adjusted, fixed, or touched-up in some way in order to pass the quality criteria.

$$First\ Pass\ Yield\ (FPY) = \frac{Number\ of\ units\ that\ exit\ a\ process\ w/o\ rework}{Total\ number\ of\ units\ that\ enter\ a\ process} \times 100$$

If units being produced have to go through multiple sequential processes, and each process has its own first-pass yield, then the overall yield of the entire group of processes is called the **rolled throughput yield**, or RTY for short. Sometimes it's referred to as the *rolling* throughput yield instead. RTY is simply the product of the FPYs of the individual processes. For example, if there are three processes in a row—Process A, B, and C—and they each have a FPY of 0.9, or 90%, then the RTY would be 0.9 × 0.9 × 0.9 = 0.729 = 72.9%. So, for a group of N processes, the RTY would be:

$$Rolled\ Throughput\ Yield\ (RTY) = FPY_1 \times FPY_2 \times \ldots \times FPY_N$$

The flip side of yield is the defect rate or the **scrap rate**. Scrap is anything that is rejected because it doesn't meet specifications. Scrap rate is determined by dividing the number of "bad" units by the total number of units produced in the same period of time. Another way to find scrap is to subtract yield from 100%.

$$Scrap\ Rate = \frac{Number\ of\ "bad"\ units\ that\ get\ scrapped}{Total\ number\ of\ units\ produced} \times 100$$
$$= 100\% - Yield$$

If you think about it, these kinds of quality metrics are actually just special cases of quantity metrics: we count units based on their relationship to some quality criteria, and then we compute a **rate** by dividing them by the total amount of units produced.

Even though I used the term "units," these quality metrics don't just apply to manufactured goods that you can touch or hold in your hand.

It's true that the fields of quality control and quality engineering often focus on manufacturing industries and production of physical goods like airplanes, automobiles, and medical devices. But the exact same concepts apply to anything made by any process that's subject to pass/fail quality criteria.

For example, a **click-through rate** (CTR) used to gauge the quality of an online advertisement or an email campaign is a form of yield. The "good" or desirable outcome is defined as a "click-through" on a hyper-link found in an ad or an email. The "bad" or failed outcome is a view (or "impression") that *doesn't* result in a click-through. So the CTR gives us the ratio of clicks to impressions.

$$Click\text{-}Through\ Rate\ (CTR) = \frac{Number\ of\ click\text{-}throughs}{Number\ of\ impressions} \times 100$$

Once we understand the concept of yield, we find this same quality ratio, often expressed as a percent, in every single industry and discipline. It's the same basic equation, just with different terms and acronyms. Human resources professionals focus on employee retention rates. Sales representatives look closely at sales conversion rates. Sports fans talk about a basketball player's free throw percentage or a hockey goaltender's save percentage. The list goes on and on. These are all quality metrics because they tell us the rate of success.

Often quality metrics go beyond rates based on binary pass/fail criteria, and the quality is expressed as a level or a grade. This can be helpful when different tiers of quality are relevant. Instead of only classifying work as "good" or "bad," it becomes useful to classify work with grades that communicate "low," "medium," and "high" levels of quality. Students become very familiar with this concept in grade school. They can pass or fail a class, but if they pass, they might have received an A, B, C, or D. Sometimes their grade includes a + or - sign (e.g. B+) to assign in-between grades.

Once we understand that these quality levels exist, we start to find them everywhere, making use of different grading scales or systems:

- Letter grades, as in grades of pearls (A, AA, AAA, AAAA)

- Named grades, as in grades of beef (USDA Select, Choice, and Prime)

- Numerical grades, as in number of hotel stars (1 to 5 stars)

Just as we moved from binary pass/fail criteria to many different grade levels, we can go a step further when determining quality and compare a measurement itself to its specification or tolerance. In manufacturing, it's common to specify **lower specification limits** (LSL) and **upper specification limits** (USL) for a measurable property or characteristic of a product. For example, a mobile device might have a minimum and maximum allowable height, width, thickness, and weight. Sometimes the specification is just one-sided, as would be the case for the minimum drop height that doesn't result in a cracked screen. You wouldn't reject a device if you had to drop it from too high up in order to get it to break. In that case, more would be fine.

This method of quality control is very effective and powerful because instead of just knowing whether or not parts meet specifications, you're able to study how well they meet the specifications, how close they are to failing, and how consistent your processes are over time. This opens the door for statistical quality control methods that we won't discuss in detail in this book.

Quality metrics are helpful when paired with quantity metrics because they act as brakes that protect against runaway speed. You don't want your team cranking out work so fast that the end product is full of mistakes and flaws. So quantity metrics are the engine and quality metrics are the brakes. They work best when coupled. If you only have brakes but no engine, the vehicle won't go anywhere. So, too, if you only stress quality, quality, quality, your team might feel overly concerned about making any mistakes, production will slow to a trickle, and you won't be able to meet the existing customer demand.

Measuring Perception

The last type of performance measurement we'll consider is quite common, and seeks to answer the question "What do others *think* about the performance?" This question can apply to the performance of people or products. And it can apply to individuals or groups.

Most often we use some form of a **Likert scale** to capture and convey perception. A Likert scale is a type of psychometric scale, which means it seeks to measure and quantify a psychological phenomenon like level of satisfaction or degree of agreement with a statement. Each one of us continually fills out surveys and questionnaires that ask us to provide our opinion about a thing or experience on a scale that looks something like this:

- Very satisfied
- Satisfied
- Neither
- Dissatisfied
- Very dissatisfied

You may use such a scale to have team members periodically rate each other's contributions in what's often called **360-degree feedback**. Or you may use such a scale in a **customer satisfaction survey** to find out how your customers feel about a product or service you provide. We all provide this kind of feedback every time we rate a book we read, a place we stay for the night, or a driver who takes us to the airport. You can provide such a rating for this very book, and if you take the course from my company, we'll definitely send you a follow-up survey to ask you to rate it. With the rise of connected mobile devices, our lives have become full of providing and giving of Likert-style scores.

It's not uncommon to turn these qualitative ratings into numerical scores: one to five stars, or "on a scale of one to ten." Whether these quantities can be summed or averaged is the source of much debate in academic circles, but practitioners do both quite often. That's why the rating for this book will have a decimal, like 4.9 stars. (You have to appreciate my optimism.)

The reason for the debate, which started many decades ago, is that these scores are actually **ordinal scales** as opposed to **interval scales**. That just means that while they belong in a specific order (5 stars is better than 4 stars, which is better than 3 stars, etc.), the distances between the values (or intervals) aren't necessarily the same. For you, there might be a huge drop in satisfaction from 5 stars to 4 stars, but for me the big drop happens from 4 to 3, or from 3 to 2. Because of that fact, star ratings are more like floors in a building, which may have different heights, than they are like cups of flour in a recipe.

Nevertheless, in practice, people are constantly converting Likert-scale feedback to quantitative values, and then summing and averaging these values to make comparisons that might not be technically legitimate, but they're undeniably useful. And so such funky math will continue to happen in spite of the purists who proclaim that it shouldn't.

A common example of an even funkier mathematical operation that people do with satisfaction scores is called the **net promoter score**, or NPS. The net promoter score is calculated from a question about how likely a customer says they are to recommend a company to a family member or a friend. If they provide a score of 0, 1, 2, 3, 4, 5, or 6, they're put into a group called "detractors." If they provide a score of 7 or 8, they're put into a group called "passives." And if they provide a score of 9 or 10, they're put into a group called "promoters." The NPS is the percent of respondents who classify as promoters minus the percent that classify as detractors. So the NPS can theoretically take any value between -100 and 100.

The trick with this form of measurement is that you have to remember that you're not actually measuring perception, but rather what people feel comfortable saying about their perception. So questions about anonymity are crucial. If you're asked to rate your boss, for example, would your score be different if you knew they'd be able to see your name attached to the score you provided? These scores are highly subjective, and many different factors can have a huge effect on the scores people provide.

You also want to avoid turning your team into a popularity contest. I want team members who feel comfortable challenging each other, and I want team members who feel comfortable challenging me. Of course I'd want such challenges to be delivered in tactful ways. But if your team members become overly concerned about how other team members will rate them, they might avoid even the kind of conflicts that need to happen in order to achieve truly high performance. How can you steer clear of this pitfall? By using a blend of different kinds of metrics, and by knowing your team members so well that you'll detect such imbalances before they tip over.

Make Metrics Relevant and Visible for Your Team

Make sure that your team members can see how well you're doing relative to your goals, and flow down all metrics to their level.

As the leader of your team, you're responsible for the overall performance of that team. As a result, you will naturally find yourself monitoring the highest-level metrics very closely. It makes sense that you'd keep an eye on those metrics, because that's your job.

One pitfall to be aware of, however, is that your team members might not feel as connected to the "big picture" goals as you do. The metrics associated with those goals might feel quite lofty and even untouchable to them.

Perhaps they feel that their specific role has only a small impact on the team's overall performance. If so, they might feel that even their best efforts won't move such a big needle. It's also possible that they don't see any connection between their own job and the team metrics. They know what they do, but don't know how their efforts affect the team metrics.

This becomes more and more likely the bigger your team gets. If you're leading a four person team like I am now, then you're all in the trenches together, working hard to move the same needle. There's actually something quite exhilarating about that. On the other hand, if you're leading an entire organization with a workforce of 40,000 employees, you have a big challenge on your hands to get them all to feel connected to the metrics.

Making Metrics Relevant

If your team members feel disconnected from or unable to affect your goals, then the metrics that you're tracking simply won't motivate them. In fact, the metrics will likely discourage them. One tool you can use to establish a firm connection between your team members and overall team metrics is called a goal tree. Figure 2.2 shows a simple goal tree of a small business looking to grow their customer base.

Goal Tree Example: Small Business Growth

Add your High-Level Goal to the top box. Specify two Critical Success Factors. Break these CSFs into lower level factors if possible. Go back and determine which are Target Variables and which are Givens, and color the headers of the boxes as indicated in the legend.

Figure 2.2. The goal tree method of flowing down metrics to more specific targets

With the goal tree, you start with one of your overall team metrics, called a **high-level goal** or **key performance indicator** (KPI). First, you articulate your goal statement related to this KPI. Then you break it down into two or more **critical success factors** (CSF). Often the CSFs can be added, subtracted, multiplied, or divided to produce the KPI, as shown in Figure 2.2. In the example goal tree, adding the number of "New Customers" to the number of "Repeat Customers" yields the total number of customers for this year.

This isn't always the case, however. Sometimes the way the CSFs combine to result in the higher-level KPI isn't quite so clean and mathematically precise. One example could be a weight-loss goal and KPI being broken down into an exercise component and a diet component. How, exactly, these two components combine to result in a decrease in body weight is difficult to know with precision. A "fuzzier" goal tree like that can still be very useful.

The next step is to break the CSFs down until you arrive at either a "Given" or a "Target." There will be some "branches" on the goal tree that you can't do anything about. For one reason or another, they're fixed, or constant. In the example above, the number of customers you sold to last year is a number you can't change. So it's either in the past, or it could also be a number that's a constraint that you simply have to accept, such as a budget or employee count figure that's locked in place for the time being.

Making Metrics Visible

Since you've gone to all that trouble to flow your high-level team metrics down to lower-level metrics, you definitely want to capitalize on that hard work and make sure your team members can see the targets that most relate to their own work. I also recommend that you show them the entire tree. That way, they can see the way they affect the high-level metrics, but they also see how someone else might be affecting them, too. So they get a deeper understanding of the way teamwork can help them all get the job done together.

You want to give each of your team members three things:

1. Visibility to the high-level metrics themselves

2. Visibility to the way the metrics flow down (via the goal tree)

3. Visibility to the lower-level metrics that they most directly affect

There are a few different ways you can provide visibility. You can show them the metrics in regular team meetings and individual meetings. The benefit to this approach is that you can have actual conversations about the metrics, and about what's going on. Some drawbacks of meetings are that they take time, it can be hard to pay attention to everything that's said in a meeting, and it can be hard to remember everything.

To get around this "fleeting" limitation of meetings, you can send a summary of the metrics (and key discussion points) in a follow-up email. The email doesn't have to be tied to a meeting, and can be sent any time—perhaps at weekly or monthly intervals. A benefit to this approach is that you can clearly articulate key points to consider in the body of the email, and it's a digital record so it's searchable. The drawback to emails is that they're static. Emails relate the facts and opinions about your metrics at a particular moment in time, and as soon as they're sent, they start to age and eventually they become obsolete.

To get around this "static" limitation of emails, you can provide access to online portals with KPI dashboards that update on a regular frequency, depending on the need. Some metrics need to be updated in "real time," but others do not. Determine an optimal update frequency for each metric and commission one or more KPI dashboards to be built for your team so that they can see the most up-to-date figures whenever they want.

Stress the Importance of Data

Make sure your team members understand how data can help you achieve your objectives, your goals, and your purpose.

As a team leader, whether you intend to or not, you will heavily influence what your team members think is important, and what they think is unimportant. You send powerful signals every time you speak, and every time you act. Even your facial features and your body language can have a huge influence on what you convey to your team. Your own internal thoughts and feelings about what really matters will come out in more ways than you will ever be aware of. So your opinion itself has a huge impact.

What's your opinion, then, about the role data plays in your team's success? The answer to this question won't necessarily be the same for everyone. For some team leaders, data is like oxygen, and the team simply can't function without it. Not even for a minute. For them, data is a resource they *must have*. For others, data is like health food. Sure, you can survive eating nothing other than junk food for a while, but you won't be in very good shape. For them, data is a resource they *should have*, but they can get by for a while without it. In theory there are teams for whom data is a resource that's merely *nice to have*, but these teams are getting more and more rare every year. You have to determine which type of situation applies to you.

Once you determine your situation, it's helpful to match the level of enthusiasm you attach to data when you talk about it to your team.

What you have is a case of *Goldilocks and the Three Bears*. In the famous fairytale, Goldilocks finds three bowls of porridge: one is too cold, one is too hot, and the third bowl is just right.

The analogy is that you can serve your team a bowl of porridge that's far too cold by **underhyping data**. If you downplay its relevance, they'll get the picture, and many of them will shy away from using it altogether. This is often the mistake of the outmoded leader who doesn't want to embrace new techniques and resources. It reminds me of the old-school baseball scouts in the best-selling book *Moneyball*, who prefer their traditional, anecdotal methods of player evaluation to the newer data-based approach pioneered by the Ivy League quants brought in to shake things up. I'll leave it to you to read the book for yourself and find out which of the two scout camps revolutionized the sport.

At the opposite extreme, you can serve your team a bowl of porridge that's far too hot by **overhyping data**. I see this all the time in my line of work. People get very enthusiastic about data and its importance, sometimes too enthusiastic. Data is pitched by big tech company sales and marketing professionals as a panacea or a cure-all. When I worked at a large business intelligence software company, I was surrounded by a lot of data pompoms, "I ❤ Data" pins, "Data Rockstar" T-shirts, and "Long Live Data!" social media post proclamations. It may have been a marketing tactic to generate harmless enthusiasm, but the message was clear: data is everything. There are lots of witty one-liners that make it seem like data is downright sacred. "*In God we trust. All others must bring data,*" said W. Edwards Deming, the so-called father of the quality movement.

The problem, however, is that data is often quite flawed. Even when the quality of the data is very high, it can sometimes be insufficient, or outdated. While the perspective data provides can be very powerful and transformational, it can also be limited in many ways. And it can even totally mislead us if we're not careful with it.

In the next chapter, we'll consider a handful of the shortcomings that your data might have. It's important to stress the importance of data, but with a healthy measure of moderation and skepticism. Data is great, don't get me wrong. But it's not a deity. Pump it up, promote it, just don't worship it. If you do, your team members will stop paying attention to what their experience tells them, and they'll stop listening to their gut.

You don't want your team members trusting their gut unconditionally either, by the way. Analytical thinking and intuitive thinking are like two airplane engines that you want firing at the same time in order to have the safest outcome. The analogy isn't a perfect one, however, because your data and your gut won't always be firing in the same direction. Sometimes they'll be pushing your team in opposite directions, in fact. In every situation, you want your team members to engage with their data as well as with their intuition, and you want to enable a healthy dialogue that brings any misalignments to the surface, so that the best perspective will prevail.

Use Data to Directly Deliver Value

Identify the unmet needs of your internal and external customers, and then find ways to use data to help them meet those needs.

For some teams, data isn't just a helpful resource to deliver products to their customers; it *is* the product. The customer can be an external or an internal customer, but the situation is similar: the value you pass along to them is data. An example of an external customer of a data product is an investment firm that purchases real-time stock market data to analyze the market and plan their moves. An example of an internal customer of a data product is a sales team that receives a spreadsheet of leads collected from contacts made at a convention or trade show.

It's possible that data is the only component of the product, as in the case of the stock market data feed. In other cases, the data is an integral part of a hybrid product, as in the case of the geospatial location data captured by a GPS-enabled watch or other device. I wear such a watch when I do activities outdoors, not just because I need to know what time it is, but because I want to know where in the world I am, and I want to record my path and my heart rate over time to review later.

In such hybrid cases, the product includes one or more non-data components (e.g. my watch) along with the complementary data (e.g.

latitude and longitude). In some situations, the customer accesses a raw data feed or extract to process, analyze on their own, and use for decisions. In other situations, the customer interacts with a visual display of the data in aggregated or summary form, such as an online dashboard or portal. Some data products allow customers to access the data in both raw and visual form.

So the data itself can take a myriad of forms, ranging from quite simple to incredibly sophisticated. Regardless of the form, though, it's important that you start by considering what *needs* the data is meeting for your customer. Of course you want your team to meet those needs as thoroughly as possible, and even exceed them if they can. You want to do so efficiently as well, because the customer won't be willing to pay for activities that don't add any value to them.

Customer Need Statements

To get this right, I recommend that you start by crafting very precise **customer need statements**. The better you understand and articulate the customer's actual need, the more likely you will be able to meet it. What is a "customer need statement?" It's a sentence, written in the voice of the customer, that clearly states what they need to be able to do or how they want to feel or be, and why that matters to them.

Here's an example customer need statement to which many of us can relate:

> *"I can more easily know when I'm at risk of spending more than I've budgeted so that I can adjust and meet my monthly savings goals."*

Let's break down this statement. If you look at it closely, you'll see that it includes five key components. The first component is the first word in the sentence, "*I.*" It orients us on who is involved. This particular statement involves the needs of an individual, but other needs statements may speak for groups of people ("*We...*").

The second component is the second word, "*can.*" This word establishes that the statement involves an enablement or an empowerment. It has a positive tone. You can also substitute this word with the phrases "*want to*" or "*need to.*"

The third component is a combination of a pair of adverbs and a verb. In this particular customer need statement, the phrase is "more easily know." The first adverb is "more," and it determines direction, so we call it a "determiner." It also modifies the second adverb, "easily," which describes how the customer wants the verb to be modified: in this case it's about the effort involved in the action. And the verb, or action, is to "*know,*" so this need relates to the customer's mental awareness of something. The verb can also be a physical action, such as "lift," or an emotional state, such as "feel."

It's possible to reword this third component in a variety of ways, such as "know with greater ease," or "know with less effort." Notice how this second variation flips the direction of the modifier from increasing ease to decreasing its opposite: effort. You can also substitute other verbs that are synonymous with "know," such as "realize," "notice," or "perceive." These alternate verbs may do a better job capturing and conveying a nuance of the action.

Let's consider other examples of the first three components of example customer need statements. We'll use fitness activity enthusiasts as our hypothetical customer. The first three components of the customer need statements are bolded. See if you can identify each of the various elements in the following:

- *"We can feel more confident that…"*
- *"I need to more accurately track my…"*
- *"I can find out in less time when…"*

In the fourth component, the customer need statement goes on to describe important details about the situation. In the example of the person watching their spending, we learn what the customer wants to know about more easily, namely, when their spending is about to surpass their budget.

The fifth and final component tells us why achieving this desired future state would matter to them—because they want to achieve their savings goal. This last component is very powerful when it includes an emotion or feeling, such as to feel less anxious or to achieve a sense of satisfaction.

Notice that the customer need statement does not simply mention a feature of your data product. "I want a spending dashboard that tells me my expenses relative to my budget." That's not a customer need, that's a product feature. It's also not a category of product attributes: "Quality," "Performance," "Price." These aren't customer needs at all.

Let's add the fourth and fifth elements of the three additional customer need statements we started above. The fourth element is shown in italics, and the fifth element is shown in bold:

- "We can feel more confident that *we have not deviated from the hiking trail* **so that we are less likely to get lost.**"

- "I need to more accurately track my *heart rate during my workout* **so that I get a good workout without putting my health at risk.**"

- "I can find out in less time when *a change in forecasted weather conditions might endanger my rock climbing excursion* **so that I can adjust my plans and stay safe.**"

How to Gather Customer Needs

In order to gather customer needs, you will need to actually listen to current customers, past customers, or potential customers. There are a variety of approaches to take, including:

- Passive methods:
  - analyzing historical customer complaint records
  - searching blogs and news articles
  - researching social media posts and online discussions
- Active methods:
  - carrying out one-on-one interviews
  - conducting focus groups
  - sending out surveys

When you are proactively reaching out to learn more, you need to ask really good questions that get to the heart of the matter. You won't identify customers' needs by asking them which of two products they would prefer, how much they're willing to pay for a product, or which feature is the most desirable to them. That's not to say there isn't a time and place for questions like those.

Upfront, however, you should ask them open-ended questions about what actions they can't currently do as well as they want to be able to do them, what ways they want to improve their lives, what emotions they feel that they want to change, and so on. A lot of times you need to peel back the layers of the onion with follow-up questions that invite them to tell you more.

For example, a hiker may tell you what they really want is "real-time weather data" delivered to their watch. That's a product feature, though, not a need. You need to dig deeper, and ask them why they want that specific feature, and what it would enable them to do that they can't do now. What are their concerns about the weather when they hike, and what kinds of decisions do they need to make?

Depending on how they answer, you might find out that a customer is only really concerned about frostbite, and if they were able to obtain an alarm for a cold-weather warning at least an hour before it hit, they'd be perfectly satisfied. You'd have transformed the statement:

- Before: *"I can get real-time weather data delivered to my watch."*
- After: *"I can more reliably predict when the weather is going to get dangerously cold so that I can prevent getting frostbite."*

What to Do with Customer Needs

Once you have gathered a set of customer needs, you need to find a way to rank them against one another. Which ones are most important, and to whom? Are there groups of different types of customers, or "personas"? They may each have their own ranked list. Surveys are useful ways of asking people which of multiple statements applies most strongly to them. There are some rather sophisticated ways of getting people to rank things, but one simple way is just to ask survey respondents to pick which three or five statements in a list of 10 to 20 most closely applies to them. Whichever need gets selected the most becomes your top-ranking need. The second-most chosen need takes the second spot on the list, and so on.

Your ranked list can then be used to brainstorm features that meet these needs. And once this list of product features has been generated, each feature can be matched with each need statement to identify important relationships between features and needs. Since each feature can either positively or negatively impact each need, this matching is best captured in a grid, with needs listed in the rows and features listed across the column headers. Arrows can be placed in each cell to indicate whether a feature helps or hinders.

Use Data to Glean Important Insights

Become proficient as a team at turning data into wisdom so that you can make better decisions that help you achieve your purpose.

Not all teams provide data directly to their customers as a product, but every team can use data to glean insights about their situation. Imagine all of your direct reports sitting around a table, but there's an empty chair. There are name tents in front of everyone, and the name tent in front of the empty chair says, "Data." Does data have a seat at the table when your team is meeting to make important decisions? You want to encourage the team to make sure that chair isn't empty. They can do that by bringing data with them to the meeting, letting it occupy that important seat, and making sure its voice is considered.

That doesn't mean, of course, that data-sourced insights are perfect. Or that they shouldn't ever be challenged by other inputs, such as experience or intuition. After all, the data you're using might be flawed in ways you don't yet see. But deciding to totally ignore the data is an approach I would recommend against in the vast majority of cases.

There are three exceptions I see to that rule of thumb. One exception arises when the data is so problematic as to be completely misleading. In a sense, you haven't totally ignored it, though, because you've learned enough about it to have the sense to avoid it. Another exception seems obvious: there may be situations where there simply is no data. Again, you're not strictly "ignoring" the data because there's nothing to ignore. A third exception concerns time.

How Much Data? How Much Time?

Perhaps the one true exception is when there is just no time to look at the data. Some decisions require split-second timing, and if you don't have a way of accessing and querying the data that quickly, you will have to proceed as best you can, and hope that the combination of intuition and luck end up winning the day for you. Those situations are relatively rare in business, I think. But in my mind, they do justify totally ignoring any high-quality data that may be available to you. After the fact, you might want to think about building a tool that can help you access the data in such a short time window.

In the final lesson of the Data Literacy Fundamentals course and book, I presented another two-by-two grid that encourages you to think about your decisions along two orthogonal axes: amount of time to decide, and amount of data available. The situation I just described is the one in the top left corner of the grid: "Lots of Data, Not Much Time."

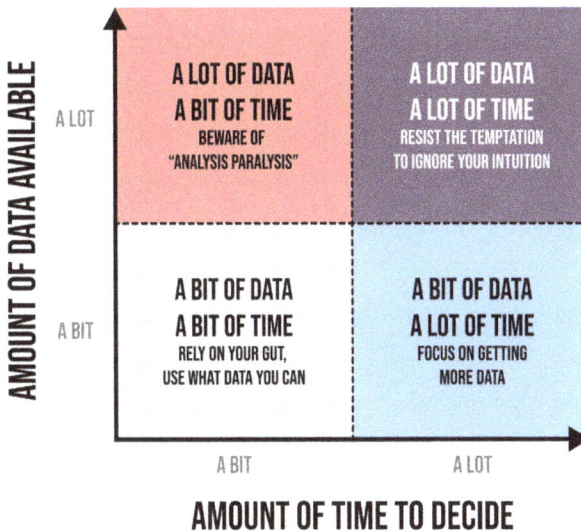

Figure 2.3. Four quadrants of scenarios relative to time and data

Turning Data Into Wisdom

Once you make the call, as a leader, that your team has enough time to consider what you can learn from data to make a better decision, the question arises: how can we get really good at doing that? In today's business environment, teams that excel at turning data into wisdom will have a huge competitive advantage over those that don't.

To get good at something, it helps to think of it as a process. What are the steps of that process, how do you progress from one step to the next, and how do you increase the fluidity of the entire chain? In today's world, I believe we focus a lot on the tools that we use to work with data, and we focus a lot on the data itself. But we don't focus very much on the process of working with data.

That's why I wrote *Read, Write, Think Data*—to encourage people to treat working with data as a process, and to help them learn how to do just that. In that book and the course that goes along with it, I describe a four-step process called the WISDOM Data-Working Flow. The four steps are Wonder, Shape, Discover, Mature, which form the acronym WSDM.

Since the Wonder phase involves bringing relevant data as *inputs* and the Discover phase produces insights as *outputs*, we have the "I" and the "O" of WISDOM: Wonder-(inputs)-Shape-Discover-(outputs)-Mature. The process is depicted as a staircase that takes us to a higher place as we progress through the steps.

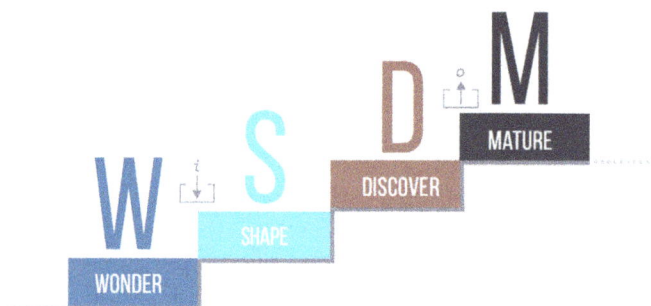

Figure 24. The WISDOM staircase analogy

In the Wonder phase, I teach readers and learners how to make keen observations, how to ask worthwhile questions, how to determine whether they have relevant data to answer those questions, and how to gather it or create it if they don't have it. The Wonder phase, as you can tell, is a philosophical one.

In the Shape phase, our attention turns to exploring the contours of the data (also known as "data profiling"), considering its potential shortcomings, finding out whether it's clean and well-structured enough, and preparing it for analysis if not. Without a doubt, the Shape phase has a technical flavor to it.

In the Discover phase, we get to dive in and analyze the data to find out what it's saying and not saying, determining whether our findings are significant, considering if we have answered our original question, and iterating if not. The Discover phase is analytical but also visual, as we think about ways to create charts and graphs that bring the data to life.

In the Mature phase, we focus on how to create a notable and memorable presentation of our findings, deliver that presentation and listen to feedback from the audience, determine if there is a need to iterate further, and then put any actions or decisions in motion. The Mature phase is all about people. This is where data gets to have its impact on the real world, through us and through our data-informed decisions.

This process has discrete steps and a prescribed progression, but the reality is that it's far from linear, it's highly iterative in nature, and it feels very messy as you're actually going through it. It's just the way it is, so that needs to be embraced. The more proficient your team can be at carrying out this process, the more value you'll extract from your data. It's that simple.

Make Strategic Decisions About Your Data

Decisions we make about data (e.g. what data to collect, how to use it) are based on an assessment of how well the alternatives will help us achieve our purpose.

It's definitely true that your data should inform your strategy. The approach you'll take to achieve your team's purpose can be shaped and guided by data that sheds light on your situation. The data can also help you evaluate different alternatives and change course when the climate changes.

But it's also true that your strategy should inform your data. What data you collect in the first place is a major statement about what matters to you. Take a look at what you're measuring, and what data you're spending your time reviewing. What do you learn about yourself? Do you have the data you need to achieve your purpose? If not, what's missing?

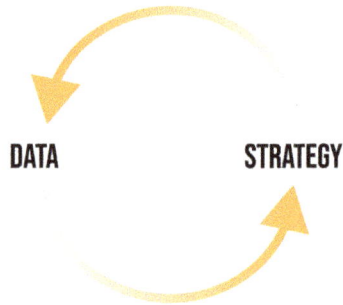

Figure 2.5. Data informs Strategy, Strategy informs Data

There are a variety of ways that your data might not be helping you carry out your team's strategy, and the tactics that flow out of that strategy. In the next chapter, we'll consider the data your team is working with in greater detail. But for now, and to bridge over into the next chapter, we'll touch on seven strategic factors to consider about your team's data:

- **Sufficiency**: Are you measuring the right things, and is the data complete enough? If not, you need to gather more of the right kind of data.

- **Freshness**: Is the data up to date compared to your team's needs, or is it getting stale or outdated? If it's stale, you need to gather more recent data.

- **Access**: Can the people who need to access the data to do their job actually access it, or are they blocked? If they're blocked, you need to get them access.

- **Governance**: Can the people who don't need access to the data to do their job or shouldn't have access because of security reasons access it anyway? If so, you'll need to block that access.

- **Meaning**: Are your team members able to come to a good understanding of what the data actually means? If not, you'll need to provide documentation or metadata to help them interpret the data accurately.

- **Accuracy**: Is the data itself accurate enough, or is it full of errors and inconsistencies? If it's highly flawed, you'll need to find ways to improve its quality.

- **Visibility**: Are your team members even aware of what data exists and is available to them, and how it can help them? If not, you'll need to find ways to improve its visibility.

Purpose Summary

As we've established, the data-savvy leader knows how to employ data in a way that moves the team closer to achieving their purpose. Instead of getting distracted by sideshows and pet projects, the highly data literate leader focuses on moving the needle.

Starting with a vision of a desired future state, the leader clearly articulates the team's purpose, and then flows that down into goals and objectives that speak to each team member. They know how to use data to inform their strategy, and vice versa. Data is a key enabler, and data-informed decisions are the rule rather than the exception.

In summary, here are seven questions for you to consider when you evaluate how you as a leader are using data to achieve your team's purpose:

1. Can you articulate your team's purpose, and are they aware of it, too?

2. What are you measuring, and how is that connected to your overall purpose?

3. Have you made your metrics meaningful and visible for each team member?

4. Are you stressing the importance of data without under or overhyping it?

5. How are you adding more value to your customers through data products?

6. Is your team adept at gleaning insights from data to inform your decisions?

7. What strategic changes do you need to make to improve your data?

Chapter 3

DATA

"No data is clean, but most is useful."

—Dean Abbott

I t may be an obvious statement, but you can't be an effective leader in the age of data if you don't get your team data that they can use to do a good job. And it follows that you can't get your team the data they need if you don't know anything about it.

The challenge is that business leaders aren't necessarily experts in data. A painter needs to be an expert in the different types of paint they use, and a chef needs to be an expert in the different types of ingredients they cook with. But does a human resources leader need to be an expert in data about their organization's employees? Does a customer service leader need to be an expert in data about their organization's call center or surveys?

A business leader doesn't need to be an expert in all of the technical aspects of their data, but they do need to be familiar enough with their team's data so they can ask the right questions about it from more technical experts on their team. In our era, data has become such an important resource for teams of all types that in order to lead effectively, you need to have your head in the data game.

In this chapter, you'll consider how you as a leader can advocate for your team, and make sure they have what they need. You'll learn to evaluate whether your data is sufficient or insufficient, whether it's fresh and timely enough, whether it's available and accessible enough, whether it's secure and protected enough, whether it's accurate and well-documented enough, and whether it's visible enough for your team so that they actually use it.

Get Your Team the Data They Need

Make sure your team members have data that's complete and sufficient enough to do their jobs well.

What data does your team need to use in order to deliver on their mission, so that they can achieve their goals and objectives and take steps toward achieving their purpose? Do they have access to that data? If they don't, how are you going to get it for them?

If you don't know how to get your team the data they need, it's very difficult to be an effective leader in today's data-saturated world. You can try to make up for this deficiency in other ways, such as having great communication skills or an uncanny ability to inspire. But the fact is that leading without data is like taking a group on a long-distance trek into the wilderness while having no idea where to find water for them.

We can separate the universe of data into four regions, as depicted in the Venn diagram shown in Figure 3.1. These regions are defined by the combinations of two binaries: 1) data you have versus data you do not have, and 2) data you need versus data you do not need. Take note of the disclaimer in the bottom right that the diagram is not shown to scale.

DATA

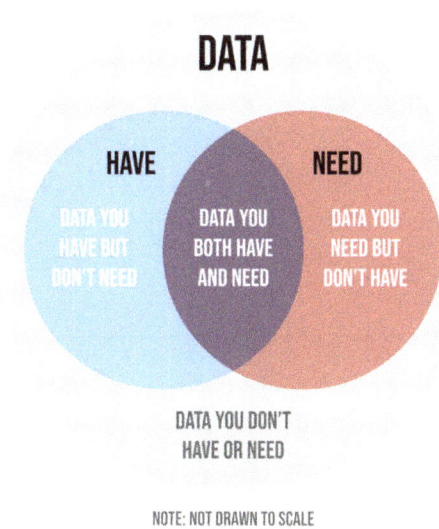

HAVE

NEED

DATA YOU HAVE BUT DON'T NEED

DATA YOU BOTH HAVE AND NEED

DATA YOU NEED BUT DON'T HAVE

DATA YOU DON'T HAVE OR NEED

NOTE: NOT DRAWN TO SCALE

Figure 3.1. A Venn diagram showing regions defined by data you have and data you need

Data You Don't Have or Need

By far the largest group is the data found outside of both circles. This is data you don't have, but that's okay because you don't need it anyway. It's mind-boggling to think about how much data is included in this region. If the shapes were sized proportional to the amount of data they contain, the rectangle would be so big that you wouldn't be able to see the circles at all. The good news is that you don't need to worry about this data, for now. You wouldn't want to spend any time with it, actually, because that would be wasted time. As soon as you've looked at data long enough to know you don't need it, you want to move on. This isn't the data you're looking for.

Data You Have but Don't Need

The area inside of the left circle but outside of the right circle represents data you have but don't need. There can be a temptation to spend time and energy working with this data in order to feel like (or look like) you're doing something advanced and sophisticated. Don't. Just pretend it isn't there.

Data You Both Need and Have

The overlapping area that's inside of both circles represents data you both have and need. Celebrate that there is good data in this region, and spend a lot of your time in it. Work with it to the best of your ability. Hire people who know what to do with it. Make sure your entire team is intimately familiar with it. Valuable data in your hands is pure gold these days. Figure out what additional ways you can extract its value.

Data You Need but Don't Have

The area inside of the right circle but outside of the left circle represents data you need but don't have. This is your job. This is where you will succeed or fail as a leader. The true test for you is whether you can go get that data for your team. Do you need to allocate funds in your budget so that you can buy it from an external source? Do you need to work with information technology to get access to it if it already exists, or gather or create it if it doesn't?

Getting into the Details

Of course there are gradations of "need," just like with patients and their medicine. Sometimes a patient will die without their medication. Other times they'll struggle mightily but they'll make it through the day. In both cases we'd say the patient needs their medication. So the Venn diagram could be redrawn with these varying levels of severity of your team's need for data. The diagram would get pretty hairy, and the point would probably be lost in the weeds. But it's important to acknowledge that you'll need to pick your battles carefully.

To further appreciate the complicated, gritty nature of reality, the truth is that most times, a particular data set will have some elements that you need and other elements that you don't need. These elements can be variables or **attributes** (think: columns in a spreadsheet), or they can be **records** (think: rows).

For example, perhaps a table of customer data includes their order history, which your team needs to understand sales trends, but it also includes their Social Security numbers, which your team doesn't need at all. Not only is this variable not needed, it's problematic because it increases the sensitivity of the data set. So the data set includes data you need and data you don't need. If you also needed shipping addresses to analyze geographic trends, but for some reason the table didn't have that variable, then you'd need to find a way to join it to the table.

The data-savvy leader doesn't necessarily need to know how to work with the data themselves, but they do need to know enough about the data that they can advocate for their team effectively. When your team has sufficient data to do what they need to do, then this part of your job is done. For the moment, anyway—the situation is always changing, and your team's data needs will continue evolving along with the changes in the environment.

Make Sure Your Team's Data Is Fresh Enough

Get your team data that's fresh and timely enough to give you an accurate picture of what's happening in your environment.

We touched on the topic of data freshness in the previous chapter on ethics when we considered how Amazon mistakenly used an outdated stock price when calculating compensation for some of their employees. Amazon is far from alone in suffering from severe indigestion after consuming data that's past its shelf life. This is a very common problem. I'd go so far as to say that every organization and every individual has suffered from it multiple times. As a leader, you'll face the problem of outdated data over and over in your career.

As Amazon's compensation calculation error proves, even the world's most influential institutions struggle with stale data. And as costly as this problem can be for employers and employees alike, it isn't always just a matter of dollars and cents. It can even cause problems when lives are on the line.

For example, in February 2022, as many governments around the world were struggling to deal with the outbreak of the Omicron variant of the COVID-19 virus, state and local health officials in California found themselves at odds with federal guidelines issued by the United State Centers for Disease Control and Prevention (CDC). At stake was whether people in different counties in California should wear masks indoors at that particular time.

According to reporting by CalMatters, a nonpartisan and nonprofit news organization, at one point in late February of that year CDC officials advised that half of the counties in California belonged in the "high-risk" tier, a classification based partly on the seven-day confirmed case rate per 100,000 residents.[40] Some officials at the county level felt otherwise, however. More recent data, they claimed, indicated that the case rate had dropped below the high-risk threshold and into the mid-tier category for many California counties, meaning indoor mask wearing should be considered unnecessary according to CDC guidelines.

The reason that the different levels of government didn't agree on the precautions that needed to be taken at a particular moment in time is that they were basing their recommendations on data from different days.

Why did this problem surface then and not earlier? During periods of relative stability of the disease, using case rates from a week ago or even a month ago might not have had an impact on the recommendations at all. But in the first few months of 2022, California experienced a dramatic spike in COVID-19 cases, followed by a precipitous decline. This surge in confirmed cases is clearly visible in the line chart in Figure 3.2. The case rate rose and then fell by a factor of 10 in a matter of mere weeks on either side of the peak in mid-January.

It's easy to see why this would make it much more difficult to coordinate efforts. The question of data freshness suddenly made an enormous difference in determining what measures were necessary to combat the virus. The data was changing very rapidly relative to how it had fluctuated in the past, so the processes and protocols suddenly strained to keep up.

Add to that the woefully archaic health data collection systems in many places around the country, and you have a gargantuan data pipeline challenge at the federal level. It's no wonder that the CDC felt more comfortable using older data than the officials at the local level were willing to use. At the federal level, the sheer size and complexity of the data collection challenge is much greater than at the local level.

40. Hwang, Kristen, and Ana B. Ibarra. "Old Data? CDC Apparently Misjudged California's COVID Risks." CalMatters, March 2, 2022. https://calmatters.org/health/coronavirus/2022/03/california-covid-risks-cdc/.

Confirmed Case Rate of COVID-19 in California Over Time

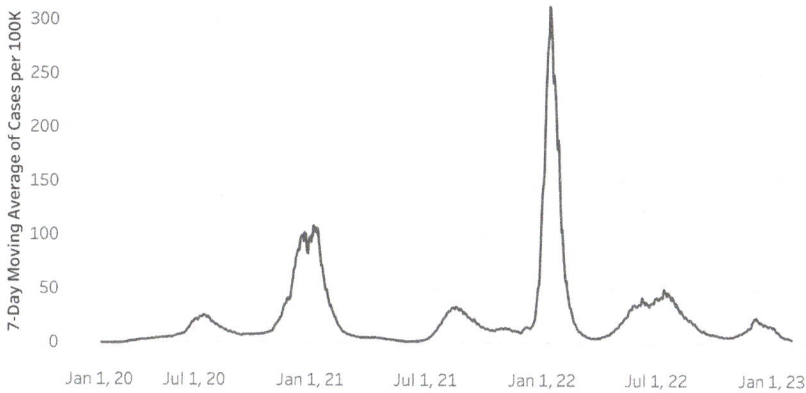

Data Source: California Health and Human Services, https://data.chhs.ca.gov/
Last Updated: February 5, 2023

Figure 3.2. California seven-day moving average COVID-19 case rate

Continuous versus Event-Based Phenomenon

As you consider the data your own team is using, it can help to first think carefully about the type of phenomenon that the data represents. Sometimes your data reflects continuous phenomena, and other times it represents event-based phenomena. Let's consider each of these one by one.

First, your data might be tracking or measuring a phenomenon that's in continual flux, like the weather, the noise level in a room, or your blood pressure. When data measures a **continuous phenomenon** like that, you want to find out how often readings are taken, called the *sampling frequency* or *sampling rate*. Are readings collected daily, hourly, or more often than that? Readings can be collected thousands of times per second, as is the case with many physical sensors such as the accelerometer in your smartphone.

Other times your data is a record of an **event-based phenomenon**, such as sales transactions, mobile phone calls, or online survey responses submitted by customers. An important consideration for event-based phenomena is how often the events themselves tend to occur. Do they happen once every few weeks, or multiple times per minute, or something in between that?

The final heat in the 100-meter dash at the summer games only occurs once every four years. That's not very often, is it? In contrast, a major online retailer might be receiving and processing hundreds or even thousands of orders per minute. When event-based phenomena occur that frequently, they can sometimes be treated as continuous.

So for data that reflects continuous phenomena, you want to ask how often *readings* occur. When data reflects event-based phenomena, you want to ask how often the *events* occur. In both cases, you also want to ask when the data set itself was last updated. And then you want to ask yourself the most important time-based question of all: how recent does the data need to be in order to support your team's processes and decisions?

Historical Data versus Current Data

Another way to think about the data set your team is using is to categorize them as either historical or current. **Historical data** is data that was collected in the past and applies to the past. Whether your data represents a continuous or an event-based phenomenon, what makes it "historical" is that whatever you were counting or measuring has changed in such a way that the data no longer reflects the present moment.

Here are a couple of examples of historical data:

- For a continuous phenomenon: the wind speed of a hurricane over the Atlantic Ocean taken six hours ago

- For an event-based phenomenon: sales transactions up through the end of the last fiscal quarter, not including sales transactions from the current quarter

When you think about the real-world phenomenon that data represents, you must remember that if your data is historical data, then by definition it isn't up to date. That is, the situation has changed in ways that your historical data does not capture or convey. This may not necessarily be a problem for you and your team. You may have a decision to make for which more recent data simply isn't necessary.

With historical data, there are two key questions to ask yourself. First, when was the last time the historical data was updated? Was it last updated yesterday? Or last year? Second, is it necessary for you to also know what happened between the last reading and now? If so, then you'll need to get more recent data. If not, then historical data will suit your purposes just fine.

If your data still reflects the present moment, then it is **current data**. This doesn't necessarily mean that the data was collected this second, or even today. Sometimes even relatively old data can be current. For example, the Winter Olympic Games are held every four years. Data from the last Winter Games might be a few years old when you use it, but it can still tell you the most recent winners in each of the events. Most people would think of data about the last Winter Games as historical rather than current. In a sense, though, it's current data until the next Winter Games is held, at which point it becomes historical if no one updates it.

When we think of current data, we more commonly think of data that updates on a very frequent basis. For example, a web analytics data set that tells you how many visitors are currently on your website isn't just current data, it's **real-time data**. Real-time data is a special case of current data: it shows up in your data set immediately upon collection, with no delay at all. Many data sets that we think of as real-time data are actually near real-time data, also called "**near-time data**." This just means that a short amount of time, typically minutes or hours, passes between a new event's occurrence and its appearance in the data set.

If we reconsider the example of the COVID-19 case rate in California during the winter of 2022, we can see that the government agencies were dealing with historical data. And the case rate for the disease was changing so rapidly at that time that it became important to get more recent data in order to make better decisions. The data collection process at the county level supported a faster update than the process at the federal level. This is why the officials at these different levels weren't on the same page. Freshness can be a tricky topic when the lineage of a data set is highly complex.

Provide Access to Those Who Need Data

Make sure your team members can access the high-quality data that they need.

If you want to be an effective leader in the age of data, you need to make sure that those who are following you are able to get access to the data they need to do their jobs. This is a very simple concept, but it's not always easy to put into practice. In my experience working with organizations around the world to assess their data literacy maturity level, one of the most common sources of frustration for employees is that they lack access to the data that they need. The larger the organization, the more common this frustration.

It's not difficult to see why employees would be frustrated by a lack of access to data. They are like the prisoner in all those movies who can see the key to their locked cell on the ground but can't quite reach far enough through the bars to get ahold of it. That feeling of being trapped and unable to break free from imposed limitations is hard enough. Add to it the tantalizing reality of a solution just out of reach and it's enough to drive a person mad.

So how do you provide access to data to those on your team who need it? Well, the first step is identifying who needs what data, and why they need it. A simple list or a grid should do fine to get started. Your IT team may use much more sophisticated tools to grant and revoke access, and you don't need to be an expert in those tools. What you do need to be an expert in, though, is what data your team needs, at least at a conceptual level, and probably at a detailed level, too.

For example, you may determine that a particular person on your team needs access to customer satisfaction survey data in order to address and alleviate common complaints. It will help you and your IT team if you are as specific as possible. Does your team member need all of the response fields, or just some of them? Would it be sufficient for them to get an anonymized version of the data, with all names, email addresses, and other Personal Identifiable Information (PII) removed from the table and replaced with a unique identifier? Does your team member need access to the responses that have been received since you started reading this sentence, or would it be good enough for them to get responses up through last month, last week, or yesterday?

Consider specifying as much as you can about your request, including the items listed below that apply to the situation. I'm not a big fan of red tape, but a basic web form or even a simple email template with these section names listed might be helpful for you.

- WHO: Your team member's name, job title, and department
- WHY: Their goal, objective, and associated task
- WHAT: The data they need, and a description of why they need it
  - The data source(s) and table(s) associated with the request

    –The fields or attributes they need

    –Any fields or attributes they don't need

  - Whether they need historical data or current data
  - The level of frequency of the data (if applicable)
  - The recency of the data (last update)

- WHEN
  - By when do they need to get access to the data?
  - For how long will they need access to it?

- ETHICS
  - Where will they store the data?
  - With whom do they plan to share the data?
  - Does the data include any personal, sensitive, or otherwise restricted information?

    –If yes, do they have the proper clearance to gain access to it?

  - Does the data have any potential for discrimination of any kind?

- OTHER DETAILS

  ◦ What tools do they plan to use to work with the data?

  ◦ Will they combine the requested data with other data?

    –If yes, provide a description of the other data and how it will be combined

  ◦ Will the data or analysis from it be published publicly externally?

If you're working in a smaller organization, or one that doesn't have a formal data access request process or form, you can still use this list to formulate your request. And you might want to provide the information in a conversational form. A more personal approach can sometimes be wise. Here is a hypothetical email that I might write to my liaison in the IT department to get data access for one of my team members:

From: Ben Jones <ben@ourcompany.com>
To: Sam in IT <sam@ourcompany.com>
Cc: Sally <sally@ourcompany.com>

Subject: REQUEST: Survey data access for Sally

Hi Sam, I hope your week is going well so far! I have a team member, Sally, who needs to get access to historical customer satisfaction survey data. She's on my product team, and responsible for improving our product packaging, which we believe is a key driver for capturing market share from Apple.

She needs to get access to the responses to the questions in the survey related to product packaging (questions 11, 12, and 13), but it would also be nice for her to compare packaging satisfaction to overall satisfaction (question 1) as well as product quality satisfaction (questions 2, 3, and 4). A full set of responses to all survey questions would be icing on the cake, but probably more than she'll use for this particular assignment.

She doesn't need any PII at all—anonymized responses with date stamps would be fine. She plans on doing a preliminary analysis of the survey data for our upcoming supplier summit at the end of March using Tableau, and combining weekly satisfaction score averages with weekly product returns data to also look for correlations there. She already has access to the product returns database. If she discovers anything insightful, she'll likely share the results in a presentation to a few of our top packaging suppliers at that summit, and they'll be getting a copy of the slides, too.

Could she get access to that data by the end of the week and through the end of the summit? Is there anything else you need from me to set this in motion? Access to the CSAT database would be ideal, but a .csv extract of the responses received in the last full 12 months would be fine, too. If you feel an extract is better, we'll plan to leave the extract on our team shared drive.

Thanks, Sam, I appreciate your help,
Ben

This is a relatively long email, but it does cover most of the items included in the list above. It probably serves as a good starting point to get Sally the access she needs. There is an advantage to this email request being in digital form; it means all of the details of the request are fully documented and searchable. There is also a disadvantage to the request being in digital form, though, because it's less personal than a face-to-face conversation.

I recommend combining the digital and the analog forms of interaction: a quick hallway conversation (analog) followed by a more detailed email request, or support ticket submission (digital). If you just submit a ticket or fire off an impersonal email, your request will simply be added to the pile. Your team deserves more than that. They deserve your proactive involvement in the interpersonal and political aspects of the organization.

Notice that the example request also involves a fairly deep understanding of what Sally is going to do with the data. In order for a leader to be able to write an email like that, they'd need to ask a lot of questions and listen carefully to the responses. There are situations in which a leader simply won't have time for that. What would a data-savvy leader do in such situations? They would ask one of their managers or a chief of staff to draft the request for them, or they would ask the person on their team who needs the data (in this case, Sally) to write it for them. Then they'd read the proposed verbiage carefully, modify it to put it in their own voice, and then send it, following up in person when the time is right.

Don't Provide Access to Sensitive Data Unless It's Necessary

Protect sensitive data by withholding access to those who don't absolutely need it.

One of the many challenges of leading a team within an organization is that your team is constantly in flux. People join your team and others leave it on an ongoing basis. Even if your team roster stays relatively constant over a period of time, team members themselves will change roles, projects, and assignments along the way.

These changes can put a very real strain on an organization's Human Resources (HR) and Information Technology (IT) procedures to onboard new employees and to make adjustments for existing ones. This is especially true of procedures to grant and revoke data access, because these procedures involve risk and they require different departments to coordinate with each other.

To kick off these procedures, you first need to determine who needs access to what data, or who needs what access revoked. Then you have to work with others to actually make that happen. With all of the constant shuffle in personnel and assignments on any given team, it may be tempting to simply give everyone access to everything.

Data for All! Free the Data! Data Democratization!

Not so fast. These flashy slogans exist for a good reason: many organizations have gone too far in restricting access to data. And while an overly rigid and closed policy can hamper progress, taking a freewheeling approach instead is a recipe for a data disaster. The corollary to the rule about granting needed access is the one about withholding unneeded access.

Sure, if everyone on your team gets access to all of the data, you won't have to worry about granting them access anymore. They may even think of you as a hero for removing all of the frustrating restrictions in their way. After all, who likes restrictions?

The problem with such a lax approach, though, is that the likelihood of a data privacy violation or a data security breach is sky high. When the inevitable happens, you'll regret not having been more careful. Guardrails are often there for a purpose: to protect you. If they're hampering your ability to get the job done, you don't need all guardrails removed, you just need different guardrails.

What I've seen over the course of my career is a pendulum effect when it comes to data access. A new government regulation or a costly data breach within an industry causes the organizations within it to clamp down on access to data. The pendulum swings to one extreme, and the policies and procedures to access data become overly restrictive. At this extreme, even employees who have a legitimate need for data that's not sensitive have difficulty getting access to it.

Eventually there's a mutiny, and the managers and employees cause the pendulum to swing in the opposite direction. The situation becomes far too permissive, setting the stage for the next data privacy or security fiasco. And the cycle repeats. We can visualize this pendulum effect as shown in Figure 3.3.

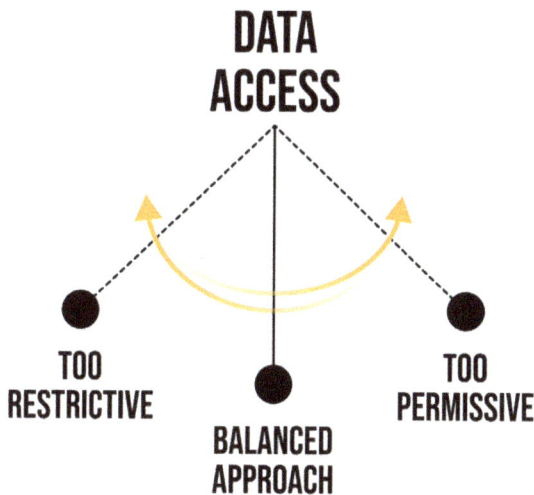

Figure 3.3 The pendulum effect of data access

Of course, what is needed is a balanced approach to data access:

- **Legitimate need for sensitive data:** Team members who have both a legitimate need and the proper clearance for sensitive data should be able to get access to it quickly, with appropriate limitations on its use. When the employee's need for the sensitive data is over, access should be quickly removed.

- **Legitimate need for non-sensitive data:** Data that has no Personal Identifiable Information (PII) and that's not sensitive in nature should be granted relatively quickly to those who need it.

- **Illegitimate need for sensitive data:** Your organization's processes should reject requests for sensitive data when the need is not substantial enough to warrant the risk associated with granting access. So you're better off not asking for such access in the first place.

In order to achieve this balanced approach, your organization will need to have an efficient and accurate process to grant and revoke data access. I've found that this very process tends to be a source of frustration for employees within organizations. We'll consider the procedures themselves in the chapter on process.

For now, your role in this process is to build strong relationships with leaders in IT. You will help yourself and your team a tremendous amount if you don't request access to sensitive data unless you can clearly demonstrate the business need to do so.

Put Good Data Documentation into Place

Make sure your most important data sets have thorough documentation so that your team members can understand what they contain, and what they mean.

Merely providing access to data isn't enough. Data by itself is just data, and it might not even make any sense to the person on your team who needs to use it. Think of the last time you encountered data in whatever form and didn't know what it meant. Perhaps there were column headers in a spreadsheet that were foreign to you. Or perhaps you came across terms or acronyms in a dashboard that you had never seen before. Maybe you saw a chart but you had no idea who created it, or when the data in it was last updated. It felt very disorienting, right?

In my book *Data Literacy Fundamentals,* I talk about the DIKW Pyramid (shown in Figure 3.4) as a way to think about the transformation of raw data to applied wisdom.

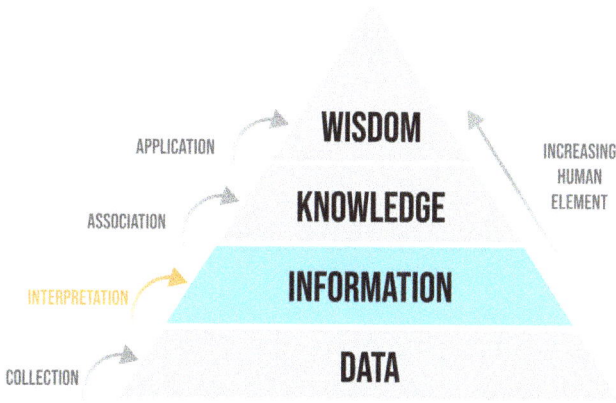

Figure 3.4. Moving from the Data level to the Information level of the DIKW Pyramid

Once data is collected, your team needs to be able to interpret it if they are to turn it into information. How can they accurately interpret it if they don't know what it means? They can make their best guess, but their guesses won't be perfect. Ascending the DIKW Pyramid requires us to add more and more of the "human element" as we go up from each level to the next. When we ascend from the "Data" level to the "Information" level, we add the human element of our assumptions about the way the data should be interpreted.

How do you increase the accuracy of your team member's interpretations of their data? One approach is to provide them with helpful metadata. Metadata is basically data about data. **Metadata** describes both the content and the context of your data. It provides useful definitions and other information that lets your team interpret it more accurately. Another way to describe metadata is as a list of relevant facts about your data.

If your team members can turn to documentation about the data you're asking them to use, then they won't have to make as many guesses, and they won't misinterpret the data as often. They'll also have the peace of mind of knowing that the documentation is there for them in case they forget, or in case they get confused. No one wants to feel incompetent in front of their peers or in front of their boss. If they have a guide to refer back to, then those very normal and very human concerns will be assuaged somewhat.

Data documentation can take many forms and it can live in many different places. Sometimes, you'll have the benefit of a fully mature data catalog owned and maintained by your IT department. A **data catalog** is a searchable inventory of all of your organization's data sources, and it's also a place to provide users with valuable metadata about each of the data sources.

Other times, you won't have the benefit of a data catalog, and you'll need to use a guerilla approach to data documentation. You and your team can work together to create your own list of common and important data definitions and you can post that list to a team shared drive. Such a manual and "DIY" approach isn't ideal, but it's probably better than nothing.

No matter where the metadata exists, you'll want it to have the following eight characteristics that form the acronym FACTS to make this list of traits easier to remember:

- **F – Fresh and Flexible:** Data can change frequently, so you want your documentation to be up to date, and you want it to be relatively easy to adjust so that it stays useful as time goes along.

- **A – Adequate and Accurate:** Your documentation should be as complete and accurate as possible, providing information about what your data contains, where it came from, when it was last updated, and who owns it.

- **C – Clear and Consistent:** The way your data is documented should remove ambiguity instead of adding to it, explaining all elements in jargon-free terms so that everyone can understand, and that's in line with the way the same elements are explained elsewhere.

- **T – Touchable and Traceable:** Your team members need to actually be able to put their hands on the documentation, and they need to be able to see how it has changed and evolved over time.

- **S – Significant and Salient:** At a minimum, your team needs to have access to metadata and documentation about the most important data sources that they need to use on a regular basis to get their jobs done.

METADATA OR GOOD DATA DOCUMENTATION

FACTS ABOUT YOUR FACTS

F RESH	**F** LEXIBLE
A DEQUATE	**A** CCURATE
C LEAR	**C** ONSISTENT
T OUCHABLE	**T** RACEABLE
S IGNIFICANT	**S** ALIENT

Figure 3.5. Characteristics of good metadata or documentation

How common is it for organizations to have good data documentation in place? I don't think it's very common at all. Even if the metadata exists, many times employees don't know that it exists, or they don't know where to go to find it. In the Data Literacy Score team-based assessment that my company runs for organizations of different types and sizes around the world, the following statement ranks 41st out of 50 statements as scored by employees of organizations seeking to assess their data literacy maturity level. This places it in the bottom 10 statements based on thousands of responses collected to date:

Our most important data sources have enough documentation and metadata to help us understand what they contain, and what they mean.

In other words, this is an area of improvement for many leaders to undertake so that their team members can feel that they have what they need to make effective use of their data. Have you addressed this common pain point yet? If not, how are you going to do that?

Strive to Improve Data Accuracy

Evaluate the quality level of your data, and, if necessary, partner with others to improve your data so that it gives your team an accurate picture of what's happening.

Data is never perfect, and your team's data is no different. It will include a whole host of errors and problems, such as faulty measurements, typos, miscategorizations, miscalculations, missing data, sampling bias, and much more. There's no point "sticking your head in the sand" and pretending these problems don't exist. If you do that, you'll be leaving your rear end sticking high in the air with a target on it. You don't want to do that.

In most cases, it's also not advisable to "cut off your nose to spite your face." In other words, just because you're dissatisfied with the state of your data, getting rid of it altogether would likely do more harm than good. Like the epigraph of this chapter states, "No data is clean, but most is useful."

So, like everything else in life, we have to work with what we have, in spite of its imperfections. As your team's leader, think of yourself as an executive chef in a restaurant, and your team members are the sous chefs and line cooks. What kind of ingredients are you giving them to prepare a great meal? Are you asking them to use stale and second-rate ingredients? That's like giving your team poor-quality data to use. If you give them poor-quality ingredients, you shouldn't expect a tasty outcome.

Assessing the Quality of Your Data

Part of your role as team leader is to make sure the raw materials that your team has to work with—including and especially their raw data—is good enough to enable them to get the job done well. You can't make much progress in this critical area unless you have an idea of the current accuracy and quality of your team's data. It's important to be aware that data is often flawed, and to be alert for errors so that you aren't caught off guard. Beyond this mindset of alertness, there are several practical methods you can employ to detect inaccuracies in your team's data:

- **By exploring the contours of your data**: Often called "data profiling," taking a close look at the attributes and details of your data can uncover many of its issues. Your goal in this step is not to answer any specific questions using your data, but rather to "size it up." Think of your data as a building or a structure and you're simply walking around its perimeter. What do you notice? Have your team members do this, but do it yourself, too. It will help you in more ways than one: exploring the contours of your data is also a great way to generate ideas about how to put your data to good use.

- **By looking at summary statistics:** If you obtain summary statistics about your quantitative variables such as their averages and standard deviations, and if you compare these statistics with the maximum and minimum values, then you'll be alerted to the presence of outliers and values that are out of the ordinary. Extreme values are not necessarily errors in your data, but they could be. Have your team find them, and then take a closer look at these special values.

- **By visualizing your data**: Nothing causes faulty values to jump out of our data quite like creating a chart out of it. Data visualization exposes patterns, outliers, and trends that are hidden in your data. When an inaccurate value or an error is present within the data, you might not notice it when looking at the raw data itself, or at a summary table of the data. Visualize it, though, and it can be impossible to miss in chart form. This approach won't necessarily expose all of the errors in your data, but it will expose many of them.

- **By reconciling your data with other sources:** On a regular basis, have your team cross-check your data with other data sources.

Do the values line up with what others are seeing? Do your calculations pass a sanity check, or are they way off? Audit your data by comparing it with what you find elsewhere and what common sense tells you that you should find.

Improving the Quality of Your Data

Once you uncover inaccuracies in your data, you need to ask yourself how egregious those errors are, and then you need to take action to fix the ones that you can't live with. You don't have time to remove 100% of the problems in your data, nor is it always possible to remove all of them anyways.

But it's not uncommon for there to be one or more glaring issues that will block you and your team from moving forward. You can live with a few thorns on a rose, but even one fly in the ointment can ruin it. You have to exercise some judgment when determining if the issue with your data is a rose's thorn or an ointment's fly.

- **Sanitation:** Sometimes you just have to roll up the sleeves and apply some elbow grease to clean your data. There are a growing number of data preparation software tools and technologies out there that can save your team a lot of time. Cleaning dirty data isn't fun, but it doesn't have to be as painful as it used to be.

- **Standardization:** You might need to convert values into a common format, such as a date format (e.g. DD/MM/YYYY versus MM/DD/YYYY) or a type of currency (e.g. $USD to euros). You may also need to change the units of measure to be based on a common scale, such as converting weight readings from pounds and ounces to kilograms. For a categorical variable like product type or country name, you might need to convert individual values like "United States," "US," and "USA" to all be the exact same value.

- **Normalization:** Data normalization involves separating large, unwieldy data sets into smaller tables with unique keys that define each record and allow you to form relationships between the tables. This is a common approach in database design, so it's likely something your IT team will have to help you with.

Normalizing your data helps to reduce the chances of redundancies and anomalies in your data.

- **Validation:** Once you have specified standard rules that certain data values should follow, you can implement validation measures to prevent someone from entering data that violates those rules in the first place. This step is proactive rather than reactive since it is designed to prevent dirty data from happening rather than correcting it. To experience what this is like for someone entering data, simply find an online form asking for your email address and leave out the @ symbol. You will get an error message that tells you that your email address is not in a valid form. As a leader, you can work to make sure that similar validation checks are in place to keep your team's data clean.

- **Collaboration:** Reach out to those who are manually entering data, managing the data, or analyzing it to brainstorm ways to improve the quality of the data going forward. You'll be surprised how many solutions you can come up with if you work together.

Working productively and proactively with your team and with others across the organization to improve the quality of your team's data is an important behavior to model. Over the course of my career, the people I have worked with who are the most highly data literate have all exhibited this behavior. They are resourceful enough to make use of what data is available, but they don't stop there. They also find ways to fix the data and improve it so that it becomes more and more valuable as time goes along.

Assign the Role of Data Curator

Figure out how your team members will become aware of important data sets as well as changes to them as they occur.

The other day my wife, Becky, and I were looking for a piece of art to hang on the living room wall of our condo in Palm Springs. At a local vintage furniture and art store, we stopped at the front desk to say hello to the owner, who also happens to be our neighbor, and we asked him what he had in the way of larger paintings. His store is beautifully laid out, with pieces arranged in a complementary way that offers a sense of how a room would feel with them in it. We walked around the store for a while admiring the different arrangements. Then we noticed a beautiful piece of artwork hanging on the wall. As much as we liked it, we didn't feel that the colors were quite right for our space.

We mentioned that to the owner, and he said not to worry; he'd take us to his warehouse, located just a few doors down. There, he told us, we'd be able to take a look at other paintings by the same artist, among others. We had visited his store a few times before, but we didn't know he had a storage space so close by. He walked us over, unlocked the door, and let us in, telling us to look around while he returned to the store to help with another customer.

The warehouse was not like the store. Instead of being arranged nicely in showroom fashion, furniture pieces were scattered around haphazardly, stacked on each other with blankets draped over them. Instead of being mounted on the walls, the artworks were on the ground and leaning up against the wall in the back of the warehouse, a half dozen paintings deep. We carefully tilted them aside one by one and looked for that perfect color combination. Alas, these weren't the paintings we were looking for.

What does any of this have to do with data? Well, your team members are like me and Becky. They're looking for something that they can't always describe, and they're not quite sure where to go to find it. To them, it can feel like your organization's data sets are in a cluttered warehouse, not an organized showroom. Like us, your team members might not be aware of what's in the warehouse, or that it's even there to begin with.

If they're as lucky as we were, they will meet a curator who will have taken the time to arrange high-value assets in a way that makes it easy to explore them. Instead of an art gallery curator or a museum curator, they need a data curator. A **data curator** is a person whose job is to collect data from different sources, organize and integrate that data, and then present it to a group who can then make use of it. As the inventory of data sets within an organization has increased, so has the importance of the role of the data curator.

A good data curator performs a number of critical functions for your team. They make sure that the team members are aware of the valuable data that exists, and they make it easier for them to find and actually use that data.

1. **Attention:** They listen to the business to understand current needs.

2. **Familiarization:** They become intimately familiar with the data that exists.

3. **Location:** They seek and find new data sets that meet business needs.

4. **Organization:** They arrange data sets and combine them together.

5. **Sanitization:** They remove sensitive aspects of the data before sharing.

6. **Documentation:** They create helpful metadata and data dictionaries.

7. **Presentation:** They put high-value data sets on display for others to see.

8. **Collaboration:** They obtain feedback on data sets and adjust accordingly.

8 ASPECTS OF DATA CURATION

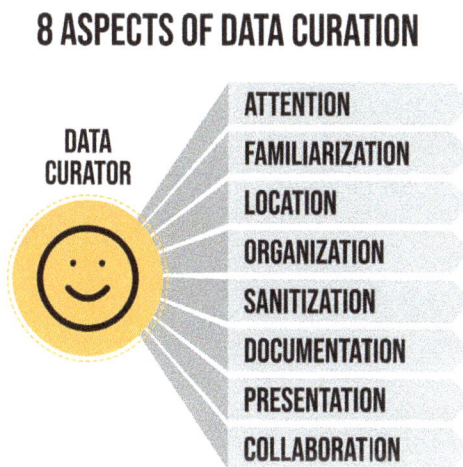

Figure 3.6. 8 Aspects of Data Curation

Data curators can use a variety of tools to do their jobs, from sophisticated data catalogs to simple data dictionaries. Beyond just using tools and technologies, though, they have an intense focus on people. They think about what people need, and they think about how to meet those needs. They have the communication skills and the presentation skills to connect with people and share information that matters in a way that resonates with them.

The role of the data curator can be an official title, or it can be an informal assignment. The data curator can work within the IT department, or they can work within the business. It's possible for a person to be an effective data curator in any of these configurations. The key is to find the person who has a unique combination of business acumen, data chops, and communication skills.

Data Summary

By now, I hope you've thought more about your leadership role as it relates to your team's data itself. Unfortunately, you can't afford to delegate the responsibility of getting your team the data they need. For better or for worse, that buck stops with you.

No, you don't need to write all the SQL queries, pivot all the spreadsheets, build all the KPI dashboards, and create all the statistical forecasts and calculations yourself. But you need to stay engaged, continue learning about your team's data, allocate resources within your team, and lobby for support from without.

More than anything, you need to ask plenty of questions about how things are going with respect to data. Here's a recap of the seven factors we considered in this chapter, this time worded in question form for you to consider.

1. How are you ensuring that your team's data is sufficient to get the job done?

2. What are you doing to make sure your data is fresh and timely enough?

3. Are you "going to bat" for your team to get them access to the data they need?

4. At the same time, are you protecting sensitive data from unneeded access?

5. Are you ensuring that your team has access to adequate data documentation?

6. What steps are you taking to improve data quality and accuracy?

7. Who have you assigned to the role of data curator?

Chapter 4

TECHNOLOGY

"Technology is cool, but you've got to use it as opposed to letting it use you."
—Prince

For thousands of years, humans have been using technology to help them work with data. The abacus was used in ancient times as a counting and calculating tool, and versions of this device have been discovered in the ancient Near East, Russia, China, and Europe. The Japanese used a similar device called the soroban, and the Incas in Peru used the quipu—a related system of knotted, colored cords to keep track of numerical data.

As ingenious as these technologies were for their time, we have moved on to technologies that are vastly more advanced and powerful, and that enable us to do magical things with data. In order for you to be an effective leader in the age of data, you need to have a certain level of familiarity with the tools and technologies that your team will need to use in order to leverage all of the data they have at their disposal.

As with the data itself, you don't need to be an expert in all of these technologies. You likely won't even need to be an advanced user of any of your team's tools yourself. Mastering all of them? That's simply out of the question. What you need is a working knowledge of the tools— what they *are*, what they let your team *do* with their data, and what that enables them to *accomplish*.

In this chapter, I'll ask you to think about your team's data tools, and the technologies that they use in order to make good use of the data you have worked hard to get for them. Do they have the tools they need? How well do they work together? Are they getting old and antiquated, are they cutting-edge, or are they somewhere in between? Do they enable you to be nimble and light on your feet, or do they weigh you down and keep you stuck in the mud? We'll consider these and other questions in this chapter on technology.

Get Your Team the Data Tools They Need

Make sure your team has the tools and technologies they need to effectively work with data to achieve their objectives.

At the risk of taking it too far, we can extend the restaurant analogy that we introduced in the previous chapter a little farther to encompass the topic of the present chapter: technology. What we've considered already is that you are like the head chef, your team members are like the sous chefs and the line cooks, and your team's data is like the ingredients that you combine to create tasty dishes for your customers.

Well, in that analogy, the knives, cutting boards, mixing bowls, pots and pans, whisks, and other instruments that the chef's crew uses in the kitchen are like the tools and technologies that your team uses to interact with and process their data in their jobs. What's the quality of those tools? Are your team's tools analogous to those of a gourmet restaurant, or are they more like those of a poorly stocked, low-budget vacation rental?

Each team has its own situation and its own needs, so no one can give you a definitive list of tools to implement that will apply to you as well as to everyone else. You can't use a "one size fits all" approach with data tools. Additionally, these technologies are evolving very quickly, so it would only take a matter of months or even weeks for a list of "go-to" data tools to become outdated. You have to assess your own team's situation, and you have to ask yourself what data tools your team needs in order to help you run your processes, make your decisions, achieve your goals, and live up to your purpose. And then you have to assess the situation all over again next quarter. Not next year—next quarter.

The least I can do, however, is to at least mention what *kind* of data tools you may need to get for your team. To do that, I'll use a framework I introduced in *Data Literacy Fundamentals* that focuses on what your team actually does with data, since we don't use tools for the sake of the tools themselves; we use them to get a job done. The framework is called "The Seven Groups of Data Activities," and the diagram in Figure 4.1 shows how they are related.

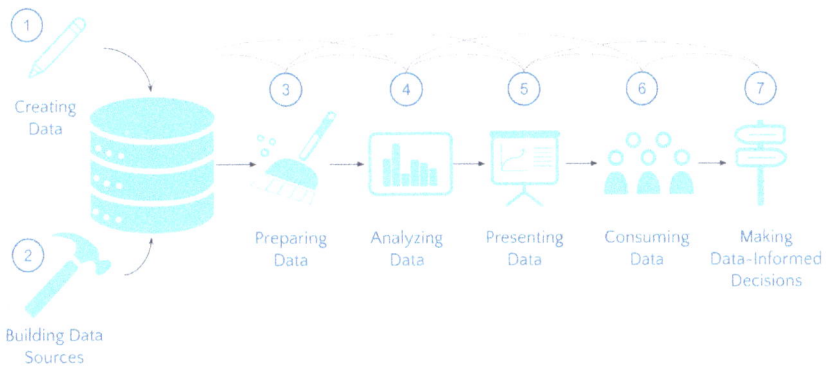

Figure 4.1. The Seven Groups of Data Activities

Creating Data

The diagram starts with creating data, and every single team creates data. Your team sends emails, interacts with mobile applications, fills out web forms, and enters values into systems such as Customer Relationship Management systems (CRM), inventory management systems, or financial reporting systems. Your team may even interact with sensors and **Internet of Things** (IoT) devices such as temperature sensors or Global Positioning System (GPS) sensors.

When implementing and fine-tuning technologies that help your team create data, the emphasis should be on three factors: accuracy, consistency, and speed. You can work with your IT department to implement tools that reduce the number of errors that your team creates, and that reduce the amount of time your team takes to enter data of various types.

Building Data Sources

Data that your organization creates gets stored in specific places where it can be accessed by people and systems that need to use it. Some data sources are very simple, such as lists and spreadsheet files that your team creates themselves and then saves to a folder on a shared drive. Other data sources are quite complex, such as **data lakes**, which can be massive repositories of unstructured data pulled together from many different sources. Unless you lead an IT team, you probably won't create your own data lake, and that's okay.

No matter the level of sophistication of the data source your team uses, you want to think carefully about how the chosen technology accommodates the required level of quality, volume, speed, retrieval, integration, privacy, and security of your data. Technologies can differ widely in these and other respects.

Generally speaking, it's normal and acceptable for ad hoc activities to result in data files and manual spreadsheets, as long as the data isn't sensitive and unprotected. On the other hand, a recurring process really should write data to a structured and well-defined **database**. Additionally, if your team will primarily be using a data source for business intelligence or reporting purposes, they'll likely turn to a **data warehouse** as their technology of choice. Data lakes and data marts are unique in that they can accommodate much larger volumes of data as well as unstructured data that lacks a predefined model, such as text, images, or video or audio recordings.

Preparing Data

Once data has been created, collected, and stored to an appropriate data source, the journey has only just begun. Once they find and get access to data, your team might need to do some work to prepare it for analysis before they're able to do anything else. The work of the "data wrangler" or "data munger" involves cleaning dirty data, pivoting data tables, combining disparate data sets together, aggregating measures (summing or averaging, for example), and more.

This work can be grungy and time consuming, but the good news is that there are modern **data preparation tools** that make this work easier than it has ever been. If your team needs to prepare data for analysis, you can save them a lot of time with a tool that specializes in these sorts of steps. Some popular data preparation tools you can find today include Alteryx, Trifacta, Tableau Prep Builder, and Microsoft Power Query.

Analyzing Data

Your team is going to need to get really good at analyzing data. I can guarantee you that. And you as the team leader are going to need to find the right tools that facilitate the type of data analysis that they need to perform. There are four main types of data analysis software packages I'd like to cover: statistical analysis tools, visual analytics tools, text analytics tools, and machine learning tools. Your team will likely need more than one type of data analysis tool.

The first type of data analysis tool is **statistical analysis tools**. These tools allow your team to perform a variety of descriptive and inferential statistical analysis on your data. This could involve, for example, running hypothesis tests on sample data such as surveys or polls. It could also involve performing regression analysis to see how strongly two quantitative variables are correlated. Some of the common statistical analysis tools in use today include Minitab, SPSS, and SAS. These tools predate most of the other tools we'll consider in this section.

Next, let's briefly consider **visual analytics tools**. These tools are sometimes called business intelligence (BI) or business analytics tools. These tools let your team create charts, graphs, maps, and dashboards out of their data sets. More than simple chart builders, they also let users perform some limited forms of statistical analysis, such as linear regression or clustering. Some common BI tools in use today include Tableau Desktop, Microsoft Power BI, and Google Looker Studio.

Next, **text analytics tools** specialize in helping your team analyze unstructured data that takes the form of text: phrases, sentences, and paragraphs. These combinations of words may come from social media posts, customer satisfaction surveys, news articles, or any other source. Users of these tools can obtain basic insights about their text such as frequently used words or common themes. They can also conduct more advanced forms of analysis such as **sentiment analysis**, also known as opinion mining, which reveals the emotional tone of a body of text. A couple common text analytics tools that people use today are Aylien and Lexalytics.

Finally, **machine learning tools**, also called data mining tools, allow data scientists to use more advanced statistical models and algorithms to make predictions or find patterns in data. What sets these algorithms apart is that they are able to learn from data to gradually improve their performance over time. That is why they belong in the realm of artificial intelligence (AI). Your team can use these powerful but dangerous tools to find hidden relationships and to classify your data. Some common data mining software tools in use today are KNIME and RapidMiner. This space will continually develop in the coming decade, and AI tools will continue emerging at an increasingly rapid pace as time goes along.

More and more, these days, data scientists are turning to programming languages instead of software packages to carry out all of the types of data analysis we've just considered, and more. Programming languages like R, Python, Julia, and C++ are popular today for data science applications. The learning curve to perform data analysis is much steeper with code than with drag-and-drop software programs, but the ceiling is also much higher. If you have a data scientist on your team who can use any of these programming languages to help you analyze your data, then you'll want to work with them closely to understand what is possible.

Presenting Data

When people present data to others within their team or their organization, they most commonly do so using everyday communication vehicles like email, messaging platforms, and the good ol' spoken word. These ways of sharing information with each other can be perfectly effective, and it's unlikely that your team will need some fancy communication technology to get their message across.

When data is presented visually, what tool or tools should they use? The most common solution is often to take screenshots of charts and paste them into PowerPoint, sharing those slides in person or on a video conference. Because this approach is so common in business, it's also widely hated. No one wants to succumb to "death by PowerPoint." But this approach can be effective enough. And contrary to popular belief, no one will kill a kitten every time a PowerPoint is made.

Your team can move beyond this basic approach and employ more sophisticated technologies such as Tableau's Stories, for which each "slide" is a data graphic that's fully interactive, rather than just a static screenshot. Another more advanced data presentation technology is the "Talkies" feature in Flourish Studio, which even lets storytellers upload an audio track to accompany their build. These more advanced data presentation technologies are like swinging for the fences: you either hit a home run or you strike out. So be careful when trying to use them.

Consuming Data

When your team wants to consume insights gleaned from data by others, what technologies can they use to do that? This is a very common use-case for teams of people who aren't highly skilled in the art of data analysis or data science. They don't have the skills to actually work with the raw data itself, so they depend on specialists outside of their team to create interfaces for them to obtain answers to their questions. There's nothing wrong with this scenario, by the way. We aren't all highly skilled automobile mechanics, either.

In those very common situations, your team will need to partner with in-house data experts to create tools for them to consume data. Most often, this takes the form of a team chart or dashboard saved to a secure location. A data dashboard might be native to an application, such as a generic Google Analytics dashboard that tells marketers how many people have been visiting a website. Or they might be custom-built for your team by business analytics professionals that work in IT or another team that services other teams or departments across the organization.

I believe we will see AI chatbots begin to take the place of dashboards in the coming decade. Instead of referencing a prebuilt dashboard that may or may not answer their specific question, your team members will simply type in their question into a chat interface and an AI algorithm will provide the answer in any of the six forms mentioned earlier. Additionally, AI assistants will generate dashboards for you based on simple prompts or questions.

Making Data-Informed Decisions

When it comes to actually making decisions, your team can leverage the entire suite of technologies at their disposal in order to inform those decisions with data. Every single tool you have just considered can potentially be a decision support tool. Some of these tools require you to be the "human in the loop," manually interacting with them to enact the decision yourself. Other AI-based tools such as recommendation engines will spit out a set of options along with a preferred path. And yet other tools will actually make the decision for you and automatically implement it. Your team's role in those more heavily automated scenarios is to monitor the decisions and processes and adjust accordingly.

Root out Any Tool Performance Problems

Make sure your team's tools perform well enough that their speed, availability, and performance don't pose a barrier to making effective use of your data.

A very frustrating situation can surface for your team members regarding their data tools: they might have access to the perfect tool for the job, but it might not be possible or feasible for them to actually use it. How can that happen? Besides the obvious issues with expired software licenses and complete hardware failures, data tools can suffer from performance issues that cause them to all but grind to a halt.

One root cause for this challenging situation is that the volume of data might dramatically change from one day to the next. As the data ramps up, what was once a perfectly snappy dashboard can show nothing but spinners for hours. An SQL query that once returned results in fractions of a second can take a very long time to execute and it can consume incredible resources in the process. Even an Excel spreadsheet with a large number of complex formulas can be sluggish or totally crash.

We've all been there before, waiting for an application to respond so that we can finish an important task we've been assigned. As the clock ticks away and the deadline looms, the sluggish program makes us want to pull our hair right out! Some data tools can really test our patience, can't they?

To find out if your team is experiencing such exasperation, you can simply ask them what mission-critical technologies are currently letting them down, and why. Chances are, though, you won't have to ask. In my experience, people are very quick to complain about such technology headaches.

It will be very rare that you yourself can alleviate the data logjam, clear the digital roadblock, or otherwise do anything about this problem, which likely has a technical root cause that only your IT department can troubleshoot and fix. What you can do, however, is document the issue. Have your team members perform a simple timing study, have them record their screen, or gather other proof points to demonstrate what's happening. Your IT team will often want to know particulars about the performance problem. What kind of hardware is being used, which operating system is your team member using, what are the system parameters or the specific details of the task they're trying to complete. Offering up these helpful clues will help your cause.

If you've gone to great lengths to get your team the data tools and technologies that they need, your job might not be done. You might need to advocate for additional support to keep these tools running smoothly. Just like analog tools, digital tools require maintenance, they're susceptible to failures and performance issues of various kinds, and they need to be taken to the shop every now and then.

You don't have to personally get involved in every such snafu, but if a trend starts to emerge, or if the problem is severe enough that your team is at risk of missing an important deadline, then you might need to run it up the flagpole, so to speak. In such moments, you'll wish you'd have built up some good will with the teams that have to step in to fix the problem.

Make It Easy for Them to Find the Data They Need

Provide your team with tools that enable them to search for and find data and visuals that they need to do their job.

In order to use data, your team members need to know where to go to find it. What technologies are you giving them to help them discover what data exists within your organization? Your team members need a map of the territory, so to speak.

When we carry out our Data Literacy Score team-based assessment, after the scoring questions we ask respondents to choose from a list of 17 common strengths and barriers. Two barriers that I have seen rising to the top of this list are the following:

- *Difficulty finding and accessing data*
- *Low awareness of what data is available*

This jives with my own personal experience working in both medium- and large-sized businesses in my career. It also matches up well with anecdotes I've heard quite often amongst my coworkers and connections in the data world: when it comes to data within an enterprise, the landscape can be downright dizzying.

Even a relatively small organization's data landscape can feel like a labyrinth to the team members who work there. A big reason for this is that data is all over the place—it's stored in different places, it's structured in different formats, and those important details aren't housed in a central location.

A technology that, when used right, can help alleviate this exact pain point is a data catalog. A **data catalog** is a central repository that contains a searchable inventory of the different data assets that an organization possesses. It also can house helpful metadata and documentation about those data sources, and it can even provide **data profiling** features that give your team members a bird's-eye view of the data. That way, after they find what they're looking for, it will take them a lot less time to find out what it actually means.

The reason that I added the qualifier "when used right" is that I've heard from organizations that have tried to implement a data catalog that it didn't always go so well. The way they implemented the data catalog didn't really help solve the problems of lack of awareness and lack of access.

Usually the symptom of the technology's failure to deliver on its promise is low utilization of the tool. What's the root cause of that symptom? Well, a data catalog, like a social networking platform, benefits or suffers from what's called the "network effect." According to this well-known effect, the platform becomes more useful as more people use it. At the beginning, though, a "chicken and egg" problem exists that those who are implementing the platform need to overcome. In order to do so, they have to apply a lot of concerted effort to add content to the platform, and they have to recruit a lot of people to start using it, and to continue using it even though they might not get a lot of value out of it right away. To do this, you need to be creative and offer them incentives, such as awarding points or badges or otherwise gamifying the experience somehow. Eventually, the value of using the platform will be obvious.

A data catalog isn't the only type of technology that you can implement to help your team members find high-quality data. It also really helps to migrate as much data as possible from disparate and isolated file folders and random repositories and into commonly used data warehouses and data lakes. It can also help to implement collaboration platforms, team wikis, knowledge-base repositories, and AI chatbots to help people find what they're looking for. It's ideal to have a small number of well-known, well-managed starting points for search that all point to a common set of high-quality data sources.

Luckily, in many cases your team won't need to access the raw data itself, and a chart or a dashboard of the data can suit their purposes. The dashboard is a type of technology that gives them access to the data through the human visual system, which is a very powerful input to the brain. Dashboards are everywhere, though, so they can get lost in the heap, just like the data itself. Additionally, a single dashboard can't answer every conceivable question that they might have. So it's important to make sure that your team members know which dashboards exist to answer which questions, and where to go to see them.

I need to add at this point, though, that technological solutions aren't the only type of solutions to the common problem of the data labyrinth.

Like any problem in any organization, it can also be addressed with people-based solutions as well as process-based solutions. We'll consider these other ways to alleviate the pain point in the next chapter. Of course, these approaches are not mutually exclusive, and the best approach is often a combination of all three.

Make Sure Your Team's Systems Work Well Together

Identify points of friction between different systems and make improvements so that the overall system feels compatible and functional to your team.

If you think of the tools your team has to work with as gears turning together in an engine, then you want those gears to fit together well, and you want them to be well oiled. This is called **interoperability**. The problem that commonly exists, though, is that these gears wear and get chipped or damaged in a way that causes all of the other gears to slip or spin or sometimes even grind to a complete halt. As true as that can be of gears in your automobile's transmission, it can also be true of the data tools and technologies your team is using.

Let's consider a couple of examples to make the analogy clearer. Let's say a particular report references a data set that gets moved, deleted, or goes offline. That report now just shows error messages where it once showed an insightful chart. Or perhaps your team needs to look up values in one application and manually copy and paste them into another system because one is a "**legacy system**" that can't talk to the newer one. Your team gets to shuttle the data over, day after day after day. The system technically works thanks to manual intervention, but you can imagine how they might object to being glorified couriers.

It's not surprising at all that these kinds of incompatibilities would surface, if you stop and think about it. Many of the software applications you use were developed by completely different teams of people, in different places, and at different times. The tools each have their own versions, and the versions your team is using might have been updated recently or a long time ago. Sometimes they employ different standards, protocols, or proprietary formats, which generates the need for a whole host of homegrown patches to get them to actually talk to each other.

As maddening as it can be, some tools are purposefully designed to be "locked in" to vendor-specific product families, so that they can't work with products outside of that family. For an everyday example of this, think back to the early days of Apple's iTunes service, which limited users to playing purchased songs on the iTunes media player itself or on an iPod. Other times, the technologies could in theory be made to work well together, but it's not a high enough priority for people to actually do the work to make that happen.

It can be a real mess under the hood. You have to feel for the folks in IT who are trying to get the whole ecosystem of tools to function. Your role as your team's leader is to identify friction as soon as it starts heating up, and to work to get things smoothed out as much as you can, and as quickly as you can. It's easier said than done, I know. If the incompatibility is particularly egregious, you may have to make concessions or allocate your own resources or budget to solve the problem. It's a give and take world, after all. You can't get something for nothing, and a team leader who is skillful knows how to navigate those exchanges.

Give Your Team Tools to Analyze Their Own Data

Make sure your team members have access to data-working tools that enable them to prepare, explore, visualize, and analyze their data.

In the majority of cases, your team can turn to prebuilt reports, charts, or dashboards to ask common, everyday questions of their data. I'm referring to the types of questions that keep coming up, week after week.

- *"How close are we to hitting our sales quota?"*
- *"Which projects are currently over budget?"*
- *"Are our website visits trending up or down?"*

As will always be the case, though, questions continually arise that aren't typical, and that are very particular to your current situation. It's possible that these questions will be highly important, but only for a very short amount of time. Their window of relevancy might be brief, and their associated opportunity very fleeting.

For these abnormal, **ad hoc questions**, it's often the case that no prebuilt dashboard will provide your team with the answer. The question might be stemming from a situation that has arisen, such as a new competitor entering a market, or a significant regulatory action or recall, or a major world event that introduces a brand new dynamic to the market. The list of hypothetical situations that could prompt a brand new question is endless. For this reason, there's simply no way to design a set of reports and dashboards that could answer every conceivable question.

As your team's leader, you need to think about how you can empower your team to handle these ad hoc queries as they arise. Do they have technologies that let them dive into the data themselves, without requiring them to possess highly specialized skills? Can they go from question to answer fast enough? If you're lucky enough to have a data analyst or even just a particularly data-savvy team member on your staff, you can task them with finding the answer.

Chances are they'll turn to whatever tool they used to build your reports and dashboards, or else they'll write a custom SQL query to pull together a single table from various sources that addresses the question of the moment. If their technical abilities aren't quite so advanced they might have to roll up their sleeves and start hacking into an Excel spreadsheet, as risky and error-prone as that can be.

In most situations, your data-savvy team member will just turn to the tool that they're most comfortable with, especially when time is of the essence. In many cases, that's perfectly acceptable, and there's no need to force them to use a tool they don't already know how to use. Whatever tool they use, they'll need to balance the competing requirements of time and quality, and they'll need to avoid making any major errors in their analysis.

The name for this model is "self-service analytics." **Self-service analytics** is an approach to business intelligence that places tools and data in the hands of the business users themselves, as opposed to leaving them exclusively in the hands of IT professionals or seasoned data experts. The promise of self-service analytics is that it will free up the business to meet their own needs with data instead of having to depend on a limited number of specialists that may not even understand the context very well.

Even though the concept of self-service analytics has been around for decades, it began to really gain momentum shortly before 2010. Companies like Tableau and Qlik breathed energy into the movement, and pumped marketing dollars into the message that business people could take the DIY ("Do It Yourself") approach to data analysis.

The message was infectious: data shouldn't be the sole domain of some "high priesthood" of controlling gatekeepers. Companies should give the data to the people! How can you disagree with a message like that? Surely you don't prefer submitting a work order to a report factory knowing that it will take them weeks to complete your ticket. Will your question even be relevant by then?

My own career coincided with the self-service analytics movement, as I joined Tableau in early 2013 and stayed there until leaving to start Data Literacy in late 2018. Over the course of my time there, I noticed that the promise of self-service analytics was appealing, but that the tools themselves didn't necessarily deliver on that promise. The fact is that tools and technologies can't deliver on the promise of self-service analytics on their own. Powerful analytics tools are necessary but not sufficient for making self-service analytics a reality. We'll consider the other critical factors in the coming chapters.

For now, here's where I'd like to leave it: if you believe that self-service analytics is the way to go, then you'll need to give your team analytics tools that will help them dive into their own data and answer their own questions. I actually believe this is a worthwhile aim, and I see self-service analytics as a potentially winning strategy. It's just important to appreciate that in order to make it happen, you'll have to put in a lot of work and overcome some substantial hurdles.

Adopt New Technologies That Will Help Your Team

Instead of being a laggard, be quick to adopt new tools and technologies that substantially improve your team's ability to make effective use of data.

Some organizations and some individuals tend to be quick to adopt new technologies, and others tend to be slow in adopting them, or resist it altogether. The reality is that in every large organization, there is a mixture of people with these extreme tendencies, and everything in between.

In his 1962 book *Diffusion of Innovations*, American sociologist Everett M. Rogers of Iowa State University wrote about the **technology adoption life cycle**.[41] This life cycle, according to Rogers, includes five different types of individuals based on how quickly they adopt a new technology. If we were to measure how long it takes individuals to do so, they'd form something akin to a bell-shaped curve, with those who take little to no time to adopt on the left, and those who take a very long time to adopt on the right, as depicted conceptually in Figure 4.2.

TECHNOLOGY ADOPTION LIFE CYCLE

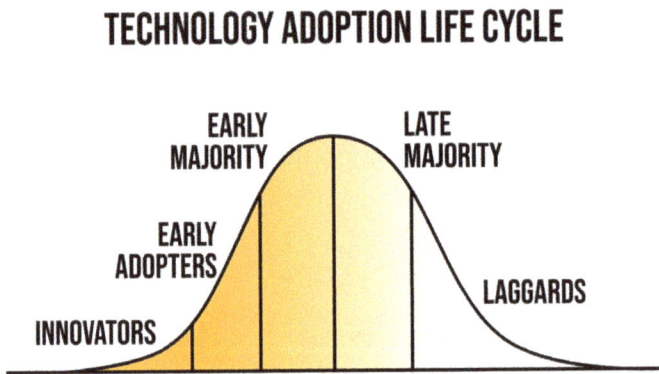

Figure 4.2. The technology adoption life cycle by Everett M. Rogers

The five groups, then, could be roughly defined by their number of standard deviations away from the top of the curve, which is the mean, or average, time to adopt. The five groups make up 100% of the population,

---

41. Rogers, E. M. (2003), *Diffusion of Innovations*, 5th ed. (New York: Free Press).

but the percentages I'll provide for each individual group are very rough figures, and can vary widely depending on the technology and the situation. We'll start from the left side of the curve and work our way to the right.

According to Rogers, the smallest group is called the **Innovators**, accounting for about 2.5% of all people, and making up the left tail of the curve—two standard deviations or more away from the mean. These individuals love to be the first to try out a new technology, and they don't mind the risks associated with using prototypes that still have many kinks to iron out. Being a "beta user" is a badge of honor for them.

People who fall into the second group, located between one and two standard deviations away from the mean on the left hand side, are called **Early Adopters**. This group makes up roughly 13.5% of the total population. Like the Innovators, the Early Adopters are tech-savvy, and they're quick to try a new technology. People in these left-most two groups either have a natural affinity for new toys and tools, or they happen to feel the pain of a problem that a new technology seeks to alleviate.

Next comes the **Early Majority**, the group that follows the Early Adopters, and they occupy the space between the mean time to adopt and one standard deviation less than the mean. This group makes up roughly 34% of the total population, and they tend to be more risk-averse, preferring to wait for the technology to become more mature and less "buggy."

Following the Early Majority, we find individuals who make up the **Late Majority**, occupying the area under the curve that lies between the mean and one standard deviation greater than the mean. Like the group to their left, this group accounts for about 34% of the total population. The Late Majority tends to be more conservative and skeptical, and people in this group are only willing to adopt a new technology after it has become well established and widely embraced by others.

The final group to the far right of the curve is the last group to adopt a new technology, and they are called the **Laggards**. Their time to adopt is two or more standard deviations greater than the mean time to adopt, and they account for roughly 16% of the total population. They are very reticent to change, and tend to be set in their ways. They may even be afraid of new tools and technologies, and are good at listing the potential negative side effects.

Which group do you think you tend to fall into? What about your team members? Do you have some people on your team who are Innovators or Early Adopters, and are there others that tend to be Late Majority or Laggards? That can be a problem, as they will tend to disagree with each other about when to adopt a new technology. Or you can view it as an opportunity: your Innovators and Early Adopters can be the beta testers for your entire team, while your Laggards can play the role of the devil's advocate and identify potential problems that could arise. Being willing to try new things while remaining alert to the potential pitfalls is a healthy and balanced approach.

In the last section, we talked about the importance of your team being able to analyze their own data themselves. Traditionally, this has involved working with raw data, preparing it for analysis, crunching the numbers using analytics software or programming languages. These are not simple tools to use, and it's easy to make mistakes with them. So self-service analytics hasn't exactly been a cakewalk for most teams.

More and more, though, I suspect that we will see an increasing number of AI-augmented tools such as chatbots and chart builders that can respond "on the fly" to questions in the form of text prompts that users provide to them. These tools will then query your data sources for you, and return an answer in verbal or visual form. These innovations will make the task of ad hoc data analytics much faster, and much more accessible to nonexperts.

The AI technologies will incorporate text analytics capabilities for summarizing and querying unstructured data such as the capabilities delivered by **Large Language Models** (LLMs). And they will also incorporate standard data analytics capabilities for interacting with structured databases to reactively return answers and proactively identify trends, patterns, and outliers.

As I write this, we are only on the cusp of this new revolution in AI-based data tools, and there are important details to resolve with the related technologies. My hope is that they are resolved in an ethical way that minimizes bias, errors, and other problematic outputs. When we place too much trust in the tool, we become the tool of the tool. We need to watch out for that. In the early stages, it will be important to at least have access to someone who can validate the answer that the AI tools provide.

Be Flexible and Adjust to the Changing Environment

Consider how the ecosystem of your team's data tools and technologies can be more flexible and adjust to the changing needs of the organization.

Being willing to adopt something new is only one form of a much broader virtue: flexibility. Being flexible and being able to adjust to a changing environment is critical, especially in industries and disciplines that are changing very rapidly. As you evaluate the technology ecosystem that your team uses, ask yourself whether or not the tools in that ecosystem are flexible. A **technology ecosystem** includes the hardware, software, and services that operate together to allow your team to get things done.

When your team's situation suddenly shifts, is it possible to quickly adjust those hardware and software tools, along with their associated services, to match the new set of needs that has arisen? Or do you find that your technologies tend to be locked in place, preventing your team from making the necessary adjustments?

CONTINUOUSLY ADJUSTING YOUR TECHNOLOGIES

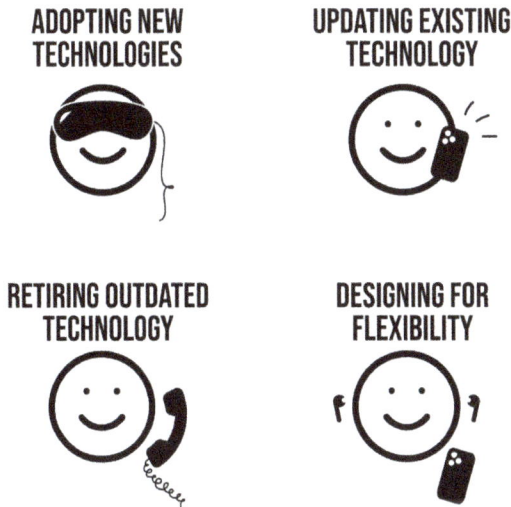

ADOPTING NEW
TECHNOLOGIES

UPDATING EXISTING
TECHNOLOGY

RETIRING OUTDATED
TECHNOLOGY

DESIGNING FOR
FLEXIBILITY

Figure 4.3. Four different ways to adapt your technologies

Adopting New Technologies

Adjusting may involve adopting and embracing new technologies, as we considered in the previous section. New technologies are constantly coming out and you will need to evaluate them and decide if and when to adopt them. But adjusting can also involve making changes to existing technologies. Software and operating systems continually need to be updated, and security patches need to be installed to resolve known vulnerabilities. In many organizations, simply updating to a more recent software version of a tool can take a lot of effort and time.

Updating Existing Technologies

Other times you will need to adjust a particular technology's settings in order to accommodate a change to a related technology or process. Giving and revoking access to data or tools based on personnel changes is one such example we've considered already. This type of adjustment needs to happen on a regular basis, even in organizations that are relatively small. But there are countless other changes that could occur that would introduce a conflict or a malfunction of some type in one or more of your technologies.

Some changes are rare but have a widespread impact. For example, the Energy Policy Act of 2005 in the United States changed when daylight savings starts and ends. Prior to the change, daylight savings started on the first Sunday in April. Ever since the change, starting in 2007, it has started on the second Sunday in March. The end date also shifted slightly. As the shift in rules approached, development and IT departments had to scramble to patch any code or system that had an internal clock.[42]

Retiring Outdated Technologies

Other times an organization needs to take a look at the technologies in their overall ecosystem, and decide when it's time to "**sunset**" an older, antiquated system. Getting rid of systems can be surprisingly difficult. There can be contractual or regulatory obligations that make it very

---

42. https://www.npr.org/2007/03/08/7779875/daylight-saving-change-is-latest-tech-worry

difficult to fully remove a technology, resulting in operational costs and maintenance burdens that drain an organization's funds and resources.

The problem of antiquated technologies that simply won't die is particularly common in the public sector. In the first year of operation of my company, Data Literacy, I signed a contract to do some work for an agency of the U.S. federal government. In order to register as a vendor of that agency, the agency required me to provide a fax number for my company. This field in the new vendor setup form was mandatory, not optional. Since I started the business in 2018 I did not, of course, have a fax number. So what did I do? I bought one online. This is a particularly ridiculous example of being forced to actually adopt an outdated technology.

Sometimes the reason that an outdated technology keeps hanging on is that the Laggards of Rogers's technology adoption life cycle of Figure 4.2 are still using it. The majority of your team members have adopted the new technology, but a handful of them decide to keep doing it the old way. This can cause all sorts of problems, from conflicting data and results to process delays and confusion galore. You probably hate to be a person who forces uniformity for the sake of uniformity. But if there are costs associated with dragging the ball and chain forward, your challenge is to find a way to incentivize the Laggards to break free from the past. I wish you luck, and I salute you!

Design for Flexibility

Ideally, your team will anticipate ways in which your technology ecosystem will need to be flexible down the road, and they will design the system to be able to adjust for the likely and the inevitable. They can't, of course, predict all of the changes that will come down the pike, so no system is foolproof in this regard.

Nevertheless, there are some wise approaches to system design that can increase the likelihood that your team will be able to make the necessary adjustments to the overall ecosystem when unforeseen changes inevitably come, such as:

- Designing **modular** components that can be easily integrated and interchanged

- Incorporating technologies that can quickly **scale** up or scale down when the demand on these systems increases or decreases by an order of magnitude.

- Using as many **open standards** and protocols as possible, as opposed to proprietary ones that tie your hands and lock you into a limited set of options

- Regularly monitoring systems, carrying out routine **maintenance**, and keeping the various technologies as up to date as possible

Technology Summary

You don't have to be a full-blown technologist in order to be an effective leader in the age of data, but you do need to be somewhat tech-savvy in order to make sure your team's tools are helping them rather than hindering them. More than anything, you need to have your eyes and ears open, and you need to ask them a lot of questions about the technologies they have at their disposal, and those that they do not have at their disposal for one reason or another.

Do note that your team members will tend to experience (and vocalize) significant pain points stemming from their technologies. It's natural to see every flaw in our tools, and we tend to remember when they don't work well, and take them for granted when they do work well. That's normal. Your role, then, is to identify and alleviate technology pain points, update and upgrade your tools as much as you can, and make sure your team members have what they need.

We considered seven different tactics you can implement to drive your technology strategy forward. We can restate these seven tactics as seven powerful questions:

1. Have you given your team the tools they need, or are some lacking?

2. What technologies aren't performing well, and how are you addressing that?

3. How are you making sure that your team members can find the data they need?

4. What are you doing to reduce the friction in your team's technology ecosystem?

5. What tools can your team use to work with and analyze their data themselves?

6. Are you adopting new tools that will help your team, or are you holding back?

7. When the environment changes, are your tools flexible enough to adjust?

Chapter 5

PEOPLE

"Train people well enough so they can leave,
treat them well enough so they don't want to."
—Sir Richard Branson

As I mentioned already, I have spent the bulk of my career so far working for companies that design, produce, and market technological solutions to very real problems. It has been rewarding for me to watch many people embrace these solutions and get a lot out of them. The feeling you get when someone benefits from something you've helped to make is a wonderful thing.

Over and over again, though, I have observed people struggle to use their new tools well, and I have even observed real harm coming to them as a result. We considered the technology adoption life cycle in the previous chapter, but that model doesn't really account for the staggering number of people who fail in their attempt to adopt a new tool. I'd argue that "failure to fully adopt" isn't just a fringe case; it's the most common outcome. Let me explain.

When we think about adopting new technologies, we tend to think about big revolutions, such as adopting the personal computer in the 1980s, or the internet in the 1990s, or social media and mobile phones in the 2000s. Sure, we adopted all of those technologies and they are commonplace tools in our world now. We don't even add them to our résumés anymore.

But think of how many times you signed up for an online account for some fancy new app and never logged back in. Not even once. Or that gadget you received as a gift over the holidays that's buried somewhere in a dark corner of your toolbox. My first job after graduating from mechanical engineering school was for a company that designed and sold fitness equipment. I remember my coworkers joking that our products were really high-priced towel racks. Amazon has entire warehouses, city blocks wide, full of once-used tools that didn't even make it to towel rack status.

All of these attempts to adopt a new tool or technology ended in failure. The same is often true of the data tools and technologies that your team has at their disposal. Utilization rates can be very low. Just ask your friends in IT: there's an entire graveyard of data dashboards that haven't been viewed in many months. The same can even be true of the tools used to make them.

In order to realize the benefits of technology, people almost always need to obtain some working knowledge and skills first. This chapter deals with the most important variable in the equation: the human variable.

Invest in Your Team's Ability to Access and Leverage Data

Make sure that the people on your team have the knowledge and skills they need in order to be able to access and leverage data.

Just like the treadmill that gets purchased but never used, or the online course that gets enrolled in but never taken, your team's data is at risk of getting collected but never analyzed. We can't just blame the tools, though. Sure, they're not perfect, and they could be better in many ways. But no matter how powerful the tools are, and no matter how rich the data is, if your team doesn't possess the traits they need to use it, then it'll all go to waste. Actually, worse than that—they might try to use it anyway, end up misusing it, and inadvertently make things worse.

The point to remember is that *people* are what move data up the levels of the DIKW Pyramid (see Figure 5.1) in order to extract value from it. Unfortunately, though, there's no guarantee that they'll do that, or that it will happen at all. What we find, in practice, is that there are a number of blockages at each level that can impede the upward flow of data toward wisdom. In this section, we'll consider some blockages that relate to the knowledge, skills, attitudes, and behaviors of the people on your team.

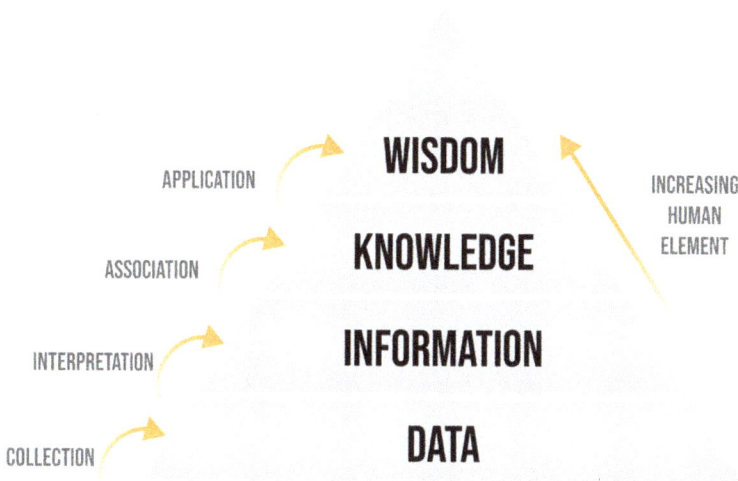

Figure 5.1 The DIKW Pyramid

Blockages at the Data Level

Sometimes the blockage is happening all the way at the bottom of the pyramid: at the level of the data itself. Data is being collected, but your team members either aren't aware of the data, or they're aware but they don't know how to find it. Let's consider each of these blockages one at a time.

AWARENESS: They don't even know what to look for in the first place

If your team members aren't aware of what data and data-working tools are available to them, then it's impossible for them to make use of what's at their disposal. I won't quote former U.S. Secretary of Defense Donald Rumsfeld's infamous and tortured 2002 statement about "unknown unknowns," as valid as it actually is.[43] But the point is that when people aren't even aware of what they don't know about, they can really stumble on their path.

---

43. "There Are Unknown Unknowns." Wikipedia, last revised March 16, 2023, last accessed May 29, 2023. https://en.wikipedia.org/wiki/There_are_unknown_unknowns.

So a basic starting point is to find ways to make sure that everyone is aware of what's in the warehouse, both in terms of data and the tools to use it. An enthusiastic and effective **data curator** can really help clear this blockage, as can a well-executed **data catalog**, both of which we touched on in earlier chapters. Whether you apply a people-based solution, a technology-based solution, or both, the idea is to make sure your team members know what high-value datasets exist, and what tools are available to them.

SEARCH: They know *what* to look for but don't know *where* to look

It's entirely possible that your team members have an idea of what's out there; they just don't know where to go to get their hands on it. Any time you've seen or heard data referenced in a presentation or a discussion but you're not sure about the source of that data, then you know how frustrating it can be. It's like that feeling we all get when we lose our keys or our wallet, and it seems like we've already looked everywhere. We really need to find it, but we have no idea where to look next.

At least if you aren't aware of what data's out there, there's a chance you'll be blissfully unaware. Unfortunately, though, what you don't know can actually hurt you. But you're not thinking about any of that; you're just cruising along, happy as a lark. As soon as you know that there's this valuable data set out there, it's downright exasperating if you don't have a clue about its source or location.

To address this common issue, your team can undergo a thorough "data orientation session." Think of your team members like people who are going to an amusement park, and there's an information kiosk just past the turnstiles where someone gives them an overview of the entire park and all of its attractions and rides, and even gives them a recommended path to take to get the most out of their day, based on their unique preferences. Who is doing that for your team and their trip to Magic Data Mountain? It's your responsibility to plan that session for everyone on your team, and to make sure that any newly hired team member gets to go through it too.

Blockages Turning Data into Information

Once you and your team members are aware of what data exists, and where to go to find it, you'll then need to understand what it means. **Interpretation** is the action you must take in order to turn raw data into information. Therefore, if you don't know how to interpret the data you've found, then you won't be able to move up the DIKW Pyramid, and you'll be stuck.

INTERPRETATION: They've found what they're looking for, but don't know how to interpret it

When you stumble on data but you don't know what it means, then you can't glean any information from it, can you? This happens quite often, and is totally normal for almost anyone who comes across a data set they've never seen before. It can be quite a disorienting feeling, and even bewildering at times.

The feeling reminds me of the "beep" noises that I used to hear coming from those toy walkie-talkies my brother Matt and I used to play with when we were kids in the 1980s. You could hold down a side button to talk, or you could push a red button on the front to send a short beep (a "dit") or a long one (a "dah") to the other person's device via radio waves.

At first, the long stream of dits and dahs were completely unintelligible to me. Then I noticed the chart on the front of the walkie-talkie: a "dit" corresponded to a dot on the chart, and a "dah" corresponded to a dash. The code for the letter A, for example, was matched with a single dot followed by a single dash: • –.

All 26 letters and 10 numerals had their own unique sequence in the chart. We worked to formulate short messages to each other that our opponents (our parents) wouldn't understand if they somehow intercepted the message. This was long before the days of mobile phones, so I must say the power in our hands felt nothing less than intoxicating.

The system on the chart was, of course, **Morse code**: the means of high-speed communication that revolutionized the world in the late 19th and early 20th centuries. When your team sees unfamiliar acronyms, metrics, or units of measure in a data set that's new to them, that's like someone hearing a Morse code message who doesn't know Morse code: they're clueless.

It pays to have your team members spend time learning the database attributes, spreadsheet columns, and terms and lingo that they'll find in the most important data sets that they need to use in their roles on your team. You want your team members to do more than merely find data; you want them to glean information from it. To do that, they need to know how to interpret the data.

MISINTERPRETATION: They've found what they're looking for, but they've misinterpreted it

There's actually an outcome that can be far worse than feeling confused by a data set that you don't know how to interpret. What happens if you feel quite confident in your ability to interpret it, but the way you're interpreting it is actually incorrect? In that case, instead of interpreting the data, you're *misinterpreting* it, and you could be misled in a way that incurs great costs to you and your team. You don't want your team members falling into this trap.

But how do you avoid this situation? One way to avoid being misled by data is to train your team members how to properly interpret their data. You can also provide them with a **data dictionary** or give them access to a **data catalog** so that they can refer to official documentation and metadata whenever questions arise.

The real challenge with this particular blockage, though, is that you might not perceive that any blockage is there at all. You feel like you're moving right up the DIKW Pyramid, from the Data level to the Information level. You wouldn't think to consult a data dictionary because you're under the impression that your interpretation of the data is perfectly fine, thank you very much. How do you build in safety nets to catch you as you fall down into the Misinformation level? How do you help your team members become aware when their interpretation of the data is incorrect?

I don't think that there's a "catchall" safety net that I can tell you to simply put in place to prevent any and all instances of misinterpretation. But I believe that one helpful approach is dialogue. The more you and your team talk about the data, the more likely you are to catch your misunderstandings and fix them.

Blockages Turning Information into Knowledge

Correctly interpreting raw data and turning it into information is not, of course, the end goal. It's just a middle step on the way up the DIKW Pyramid. What you have at this level is simply a single piece of information—a dot—floating around in your head. Well, the way to continue to step upward to a higher level on the pyramid is to make powerful associations that form connections between these individual dots.

This is what's meant by turning information into knowledge. We grow in knowledge when we connect one piece of information to another. The connection makes the current flow, and the light bulb goes on. And once again, blockages can prevent this transformation from taking place. Let's consider two such blockages.

ASSOCIATION: They aren't engaging the parts of their minds that suggest associations

American virologist Jonas Salk, who is widely credited with developing one of the first successful polio vaccines, once said, "Intuition will tell the thinking mind where to look next." But a common misconception about data is that it replaces a person's intuition. Modern research into behavioral psychology and cognitive science suggests that our brains actually engage in two parallel modes of thinking: reasoning and intuition. This theory, called **dual process theory**, contradicts the popular myth about data, and it would seek to restore our intuitive mind to its proper seat of honor. Our intuitive mind, then, is just as valuable, and just as flawed, as our analytical mind.

The best approach is to encourage your team to engage in both modes of thinking at the same time. When they interact with data, they need to learn to listen to their intuition about what they're seeing, and they need to pay close attention to their gut about where to look next. If they engage with these more instinctive, creative, and spontaneous thoughts, they're more likely to come up with powerful associations that will unlock the true value in your data. If they just sit there and buy into the false narrative that data replaces and obviates their intuition, then they won't give themselves license to dream a little.

You don't want that. You hired people with experience, ingenuity, and flair. You want them to incorporate those gifts into their analysis. I don't mean to say, by the way, that they should trust their gut unquestioningly. Like data, this can sometimes lead them astray. But ignoring it altogether is a surefire way to miss the many helpful clues and pointers that it provides. By bringing their intuition into their analysis, they add the "human element" to the chemical reaction that occurs in their brains when they interact with data.

COMPARISON: They're only able to find part of what they need to connect the dots

Just because your team members dream up a potentially powerful comparison doesn't mean that they'll necessarily have the ability to make that comparison. An inability to connect the dots is another kind of blockage that can prevent your team from moving from the Information level up to the Knowledge level of the DIKW Pyramid. What does this look like in real terms?

Imagine that you're in a meeting with your team, and you're having a discussion about an important data set—let's say sales by product for the past fall season. Someone on your team notices that sales for a certain product spiked in September, and they hypothesize that the reason for the surge could be that college students returned to campus for the fall term that month, leading to the spike.

This is just a hypothesis at this point. It may or may not seem plausible, but you only have one piece of information: historical sales from the fall season. How do you connect the dots to turn information into knowledge? To begin to test the hypothesis, you might want to look for correlations: how did last fall's sales compare with previous year's fall numbers? Is there a broader seasonal pattern? What about the demographics of customers—were their ages consistent with college students? Did sales in large college towns surge more than other places?

The blockage manifests itself when you look around at each other and discover that no one has access to these other pieces of information, or the underlying data behind them. Your team member's intuition proposed a fantastic hypothesis, but you can't test the hypothesis because your information is limited. That can be a very frustrating experience.

We'll save the last leap—the leap from the Knowledge level to the

Wisdom level—for later in this chapter. For now, I'll simply state that people need to be empowered to apply their knowledge. And they need to feel that they have input to important decisions. Satisfaction comes from applying oneself and making a difference in a meaningful cause. If you don't let them do that, then what good are their data skills?

Build Your Team's Fluency in the Visual Language of Data

Make sure that the people on your team can read and interpret visual displays of data well enough to understand the present situation accurately.

It's very important to be able to read and interpret graphical displays of data in our world. It's so important, I could write an entire book about it. That's why I did. It's called *Learning to See Data,* and it's the basis for my company's Data Literacy Level 1 course.[44] The fact is that your team members will constantly encounter charts, maps, and dashboards, and they'll be asked to use them to evaluate their environment and to make better decisions.

They'll need to be able to apply this skill at home and within their communities too, not just at work. Whether they're seeking to understand election results or stay abreast of a rapidly spreading pandemic, they'll need to get comfortable reading charts of all kinds. In 2020, as the COVID-19 pandemic overtook the world, we were presented with charts and maps in the news and on social media, and we struggled to understand what they meant. It became a matter of life and death as we sought to use what information we could to keep our families safe and healthy.

Your team likely won't ever be faced with such grave and dire consequences at work, but their ability to quickly identify trends and patterns in data can be the difference between success and failure for your team and your organization. How do you make sure your team members are fluent in this critical language?

44. "Data Literacy Level 1." Data Literacy, January 24, 2023. https://dataliteracy.com/data-literacy-level-1/.

There's no doubt that formal training courses can sometimes be help-ful, and you can always buy them books like mine on the topic. But the best way to build their fluency in the visual language of data is to actually immerse them in the language. You do that by continuously exposing them to charts and graphs of your most important data sets. And you capitalize on this exposure by fostering a healthy dialogue about those charts.

Encourage them to have open conversations about what the charts in front of them say, and what they don't say. Here are just a handful of good open-ended questions to jumpstart the conversation:

- What does the chart tell us about our situation?

- What patterns, trends, or outliers do we notice?

- What questions can we answer using the chart?

- What questions can we not answer using the chart?

- What elements of the chart are enlightening?

- What elements of the chart are confusing?

- What new questions emerge based on what we see?

It's critical for you to make it clear to your team members that they don't need to have all the answers, and it's okay to express confusion about a chart, a data value, or anything else for that matter. It can be quite scary to admit to your peers and your boss that you don't know something. For many of us, a traditional public school education has trained us to give nothing but correct answers. But when a team is exploring data together, what they really need are good questions.

Keep in mind that every chart is an encoded message. Unlike a secret code, a chart is designed to be as easy to decode by as many people as pos-sible. A chart represents your data using **graphical marks** like bars, dots (points), lines, or areas such as pie slices. The way a chart encodes your data is by using what's in your data to control the properties of those marks via **visual encoding channels** such as position, size, length, color, and more.

For example, the graphical marks used by a bar chart are its bars, and the visual encoding channel it uses is the length of those bars (see Figure 5.2). The chart features a different bar for each level of some categorical variable in your data (say, fruit type), and the length of those bars encodes some quantity of those fruit (maybe the total weight or average price of each fruit type).

Technically, a bar chart also encodes your data using two more visual encoding channels: the position of the edges of the bars as well as the two-dimensional area of the bars, which are rectangles having an area equal to the product of their length and their width. This is one of the reasons why bar charts are so useful: they give the human visual cortex three different ways to compare the values they encode.

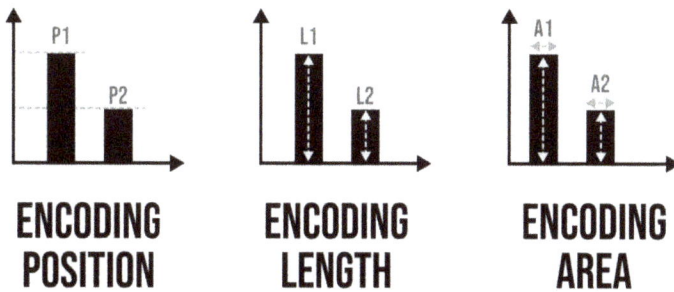

Figure 5.2. The three visual channels a bar chart uses to encode its values

There are two important principles to keep in mind when evaluating your team's charts. They are known as the expressiveness principle and the effectiveness principle. These principles form relatively loose guide-lines, as opposed to rules that are written in stone. Let's consider both of these principles, and as we do so, I'd like to encourage you to look at your team's most important charts, and ask yourself how well they follow these principles.

Expressiveness

The **effectiveness principle** states that your team's charts should express what's in your data, and they shouldn't express what's *not* in your data. That sounds like it makes sense. But wait a minute, how can a chart express something that's not in your data?

Unfortunately, there are many ways it could do so. To introduce one common way that a chart can violate the expressiveness principle, let's first think about the variable we just discussed, fruit type. Let's say there are three types of fruit in your grocery cart: apples, bananas, and oranges. Fruit type is a **nominal scale variable**, meaning the levels of that categorical variable don't belong in any inherent order. Which is the greatest fruit type and which is the worst? You may have your opinion based on your own personal preference, but the definitions themselves don't put one fruit type *inherently* above any other.

Since the fruit type variable is nominal, you wouldn't want your chart to give the reader the wrong impression that the levels belong in some inherent order. How could your chart do that? Well, if you used successively darker shades of gray to encode the fruit types, this choice would express that they belong in an order, which would be misleading.

Alternatively, if you encoded meat grade, an **ordinal scale variable**, using gray (or any color) saturation, then this choice would express the inherent order in that categorical variable quite well. Consider the two hypothetical chart legends shown in Figure 5.3, and you will see for yourself which one expresses what's in the data, and which one expresses what's not in the data.

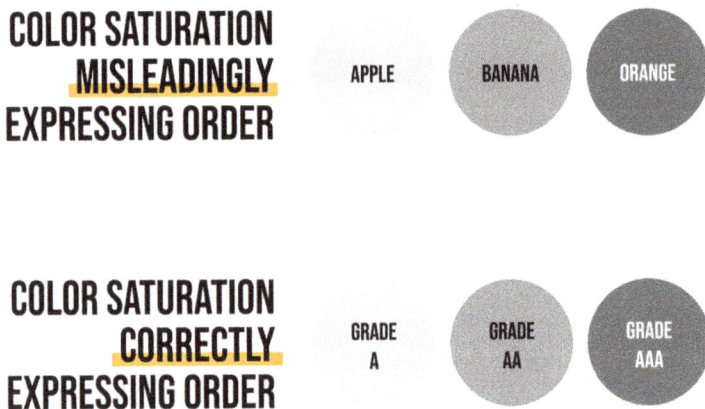

COLOR SATURATION MISLEADINGLY EXPRESSING ORDER

APPLE   BANANA   ORANGE

COLOR SATURATION CORRECTLY EXPRESSING ORDER

GRADE A   GRADE AA   GRADE AAA

Figure 5.3. Using an encoding for a nominal variable that doesn't express order

Deciding whether to encode a variable in a way that either expresses order or doesn't express order is only one choice that affects whether a chart follows the expressiveness principle, though. Different chart types tend to express different kinds of relationships in the data. When we look at a certain chart type, we tend to want to make certain kinds of comparisons, but not other kinds of comparisons.

Let me give you a few examples to help make this point clearer (see Figure 5.4). A bar chart tends to suggest to our minds that we should be comparing one bar versus another as discrete comparisons: this versus that. A line chart tends to signal to our brains that something might be trending up or down or changing over time: then versus now. And a pie chart, by its arrangement of the various slices into a complete circle, seems to say to us that we're looking at distinct parts that add up to some whole amount: part versus whole.

BARS   LINES   PIE SLICES

EXPRESSES DISCRETE COMPARISONS   **EXPRESSES TRENDS OVER TIME**   **EXPRESSES PART-TO-WHOLE**

Figure 5.4. 3 common chart types and the relationships they tend to express

Believe it or not, bar charts, line charts, and pie charts together cover the majority of cases, as long as we consider the bar chart to have either horizontal bars or vertical ones, the latter sometimes being called a "column chart." If we add to these three chart types—the scatter plot, the symbol map, and the heat map—then we will have rounded out the top six chart types, covering more than 80–90% of cases, in my experience.

Of course there are hundreds and hundreds of other types of charts, from bubble charts to waterfall charts to area charts to Gantt charts to Sankey diagrams. We can't cover all of the many chart types in this chapter, nor is it necessarily your job as team leader to be intimately familiar with each and every one of them. But what I'd suggest that you do is engage with your team members in a discussion about what your charts seem to express to each of you. And then dive deeper and ask whether this is an accurate expression and a helpful one, or a misleading or confusing one.

Effectiveness

The second principle to help you and your team evaluate your most important charts is the **effectiveness principle**. Before I define this principle, let's think broadly about the concept of effectiveness. How do you know whether a particular chart is "effective" or not? Well, you would need to start by defining what the purpose of that chart is, right? If the aim is to grab a distracted reader's attention, then has the chart fulfilled that purpose? How do you know? If the aim is to help a person or a group of people perform a task well, then does it do that? Again, how do you know?

Charts can have multiple purposes, and sometimes one person's purpose in creating it or reading it is different from another person's purpose in doing so. For one person, a chart may be highly effective, while the same chart might not be effective at all for another person. So in a general way, you can assess the effectiveness of a chart by asking whether it has hit the mark, whatever mark you were aiming for, provided it's an ethical one.

There's a more particular way that we can gauge a chart's effectiveness, though. Most commonly, when a team of people are using a chart at work, that chart's primary purpose is to help them answer a question, carry out a task, or make a decision. To do this effectively, the chart needs to enable the people on the team to make comparisons about their data, and it must:

- Have some level of **accuracy** or better
- Happen within some amount of **time**
- Require some level of **effort** or less

The accuracy, time, and effort limits can be hard to measure, though, and they can vary wildly for different tasks. Does someone need to know the change in revenue to the nearest tenth of a percentage point, or do they just need to have a rough idea, plus or minus a few percentage points, for example? This is analogous to the difference between a toy airplane that has very loose engineering tolerances for its various pieces, and a real airplane that has very tight tolerances for many of its manufactured parts. Similarly, the need for precision in comparing relative values in data can be high, low, or somewhere in between.

The **effectiveness principle** states that the most important variables should, in general, be encoded using visual channels that enable the chart reader to make more accurate guesses about the relative values in the data. That's a mouthful, so I prefer to give an example. Consider the following four charts that use different visual channels to encode the exact same four values, A, B, C, and D. If you use each of these four charts to estimate how much bigger the mark for C is than the mark for B, which encoding would give you the most confidence in your guess?

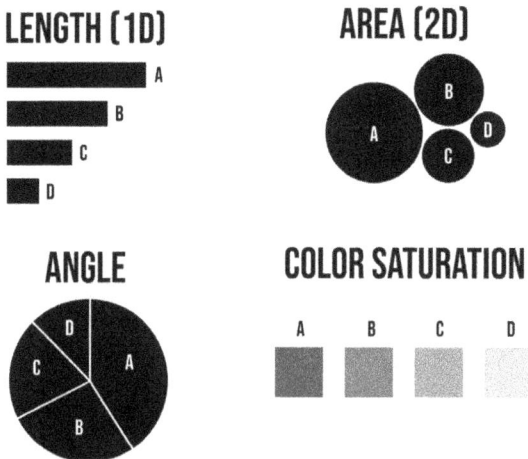

Figure 5.5. Comparing mark sizes using various encoding channels

The answer is that C is a little more than one and a half times the size of B. To be more precise, C is 1.56 times as big as B, or 56% bigger. More importantly, though, research over the past few decades has shown that people tend to guess more accurately when they are shown the bars as compared to the other three encoding types shown in Figure 5.5. The following list gives common visual encoding channels in order of decreasing accuracy:

- Position as compared to a common baseline
- Position as compared to a different baseline
- Length (one-dimensional)
- Angle (or tilt)
- Area (two-dimensional)
- Color saturation or luminance
- Volume (three dimensional)

If we were to rigidly adhere to this principle, then we'd only ever use dot plots and bar charts, which encode data using position and length. Clearly that's not the world that we want to create for ourselves. Instead of such a bland and uniform world, we want to allow for a colorful and creative world with variety. After all, a chart's purpose might be to inspire us or to give us joy.

That being said, I believe it can be helpful to think about this variation in the accuracy levels that different encodings afford us, and to at least consider this narrower perspective about a chart's effectiveness.

For highly crucial presentations of data, such as for an important annual meeting, it can be a good idea to hire or bring in a data visualization expert to provide feedback on your charts and to help you improve them. But for everyday uses, just engage your own team in a dialogue about your charts, and see what comes out of the conversation. The more fluent you all become in the visual language of data, the more value you'll get from your data.

Enable Your Team Members to Dive into Data Themselves

Tap into the knowledge and skills necessary to clean, explore and analyze raw data sets so that your team can discover important findings within your data.

The majority of individuals on teams these days take in data through facts and figures that they read, or through simple tables, basic charts, and data dashboards that they encounter as they go about their day. These presentations of data are most often created by others, and shared with them via email, on various instant messaging platforms, or on internal shared websites or portals. For these scenarios, your team members need to understand the fundamental aspects of the data being displayed, and they need to be able to think critically about what they're seeing.

These tasks are sometimes referred to as tasks of the "data consumer." I admit that I'm not a huge fan of this term, mostly because I feel that our society puts too much of a focus on consuming things. If, however, I put aside my personal distaste for this term, I admit that it generally makes sense to people. "Data consumers" aren't processing raw data themselves, nor are they diving into the data and conducting their own in depth analysis. They're processing insights that other people have found in the data. We sometimes refer to this role as the role of the "data citizen" instead, though this term is also imperfect for its own reasons.

Every now and then, though, your team's existing reports just won't answer a critical question that arises based on unique circumstances. These situations call for "ad hoc analytics." In such moments, your team will need to find a way to harness the complementary skills of data preparation and data analytics. Whether they do it themselves or tap into another resource that can quickly understand their need and do it for them, you will need people who can prepare and analyze data.

Data Preparation

When your team works with raw data tables to use them in ways they've never used them before, they often need to transform the data before they can actually analyze it. This process is sometimes called "data wrangling," or "data munging," and it can be very time-consuming and downright ugly. Besides finding and cleaning typos, missing values, and errors of other kinds within the data, they often need to restructure data tables and combine them with other data tables. (See Figure 5.6.)

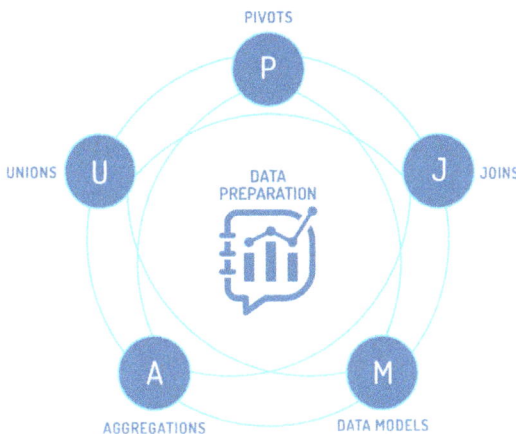

Figure 5.6. Five Forms of Data Preparation

Sometimes your team will need to **aggregate** the values in a table. For example, they might have a table of a large number of individual sales transactions, and they need to find sales by week, by month, or by quarter. To do this, they can create a summary table that sums up sales amounts by different time periods.

Time isn't the only variable they can use to aggregate their data, and summing values isn't the only way to aggregate them. Perhaps they're looking at employee salaries, and they'd like to compare average salaries of different teams or departments. They could aggregate the data by creating a summary table of average salaries of either one of these organizational structure levels. In this way, they transform granular data into summary data by aggregating it into groups that are meaningful for their analysis.

Pivoting a table involves converting a table's columns to rows, or vice versa. For example, if they have a table of financial figures that places each month's sales in a separate column—one for January, one for February, and so on—then they might need to "pivot" those columns and place them all in a single column ("Month") with all of their corresponding sales values in a single column next to it ("Sales"). But why would they need to do this? Well, the tool they're using might make it much easier to work with data in one format over the other. Or perhaps they want to display the data in a table that's structured to be easier for people to read.

Often they'll find that a data set they're working with contains some of the data but not all. For example, maybe their website traffic data tells them the country each pageview came from, but it doesn't tell them the population of that country. If they want to calculate a "visits per capita" metric, then they'll need to find a way to **join** website visits by country with country population data. You can think of joining tables as bringing their columns together and forming a single, wider table. Website visits from each country get placed next to that country's population in a single table, and these values are divided to find the view count **normalized** by population.

Other times, the data they want to analyze will be in separate tables that are otherwise identical. For example, if they want to analyze sales for different states in Australia, but the data has been broken up into eight different tables—six tables corresponding to the six states of Australia and two corresponding to the two territories—then they'll need to perform a **union** to bring them all together into one table. You can think of a union as a way to create a taller table that has been formed by combining the rows of multiple tables. Joins combine columns together to form wider tables; unions combine rows together to form taller tables.

A way to handle all of these steps at the same time is to create a **data model**. A data model allows data workers to create relationships between tables so that they can query them together. These steps and others, such as filtering, sorting, splitting, and transposing, make up the "T" for "transform" in **ETL**—Extract, Transform, and Load. We won't cover each of these possible transformations, but suffice it to say you will find times when your team needs someone who is adept at working with data in this way.

Data Analytics

It goes without saying that the purpose of working with data is not to simply transform it for the sake of transforming it. As wildly enjoyable as it can be, data preparation is a means to an end. It's the answers that surface from data analysis that your team is really looking for. What kinds of answers are they looking for? In *Data Literacy Fundamentals*, we covered the Five Forms of Data Analysis shown in Figure 5.7.

1. Descriptive	2. Inferential	3. Diagnostic	4. Predictive	5. Prescriptive
"What happened?"	"What about the rest?"	"What's going on under the surface?	"What is likely to happen next?"	"What should we do about it?"

Figure 5.7. The Five Forms of Data Analysis

As you can see, these five forms of analysis address different questions that your team can ask of your data. With descriptive statistical analysis, your team is simply trying to find out *what happened* in the past?

Inferential statistical analysis involves the special case of sample data, when your team has data from a portion of a population instead of the entire population. In such cases, they are seeking to ascertain what they can reasonably infer about the population from the sample: *what about the rest?* Diagnostic data analysis is like a doctor performing more in depth analysis to go beyond the symptoms and to find out *what's going on under the surface?* Predictive and prescriptive analysis both deal with the future: *what's likely to happen next?* and *what should we do about it?*

Knowing which of these forms of analysis is appropriate in a given situation, and then actually knowing how to carry out that form of analysis are two different skills. Chances are, you'll need to find ways to tap into both skill sets. Who will help your team with these tasks when the need arises? You either need to 1) hire someone who can do these tasks, 2) train someone to be able to do them, or 3) find a shared resource who can do them for you. Once you find that person or those people, take good care of them. They're priceless resources.

Turn Your Team into Great Data Communicators

Invest in your team's ability to visualize, present, and communicate data findings well enough to accurately inform others about what's happening.

Like the proverbial tree that falls in the forest with no one around to hear the sound, we can ask whether a data insight that's never communicated to anyone has any impact at all. Sure, the insight might have an impact on the one individual who discovered it. But if that's the end of the story, then it has fallen far short of achieving its full potential to drive change.

In order for an insight to have a significant impact on an organization, it almost always needs to be communicated to others first. The reason is that people need to understand the finding so that they can get on board and help set things in motion. They probably need to push back and talk about their concerns first, too. Change isn't easy: adjustments will need to be made; funds will need to be budgeted; resources will need to be allocated. To make that happen, your team will need to become talented data communicators.

One challenge is that communicating with people is a very different skill than exploring data. It involves a totally different mindset. They will need to shift gears from the analytical mind to the interpersonal mind. What triggers this gear shift is empathy. When your team can empathize with their fellow team members, their peers, their coworkers—their audience—then they can imagine what it would be like to be them. When they can imagine what it would be like to be in their audience members' shoes, then they can craft a message that their audience will be more likely to hear and receive.

What form of data do your team members share, though? In other words, when they present data, what does the audience actually *see*? The presenter can't just point to the entire database and say, "See, there you go!" Of course no one would actually do that. That would be rather extreme. Unfortunately, the opposite extreme—brilliant data communication—is quite rare, too.

In *Data Literacy Fundamentals*, I attempted to answer this question by outlining "The Six Ways of Displaying Data." When your team communicates data, they share either 1) facts and figures, 2) summary statistics, 3) tables of numbers, 4) charts and graphs, 5) dashboards with multiple charts, or 6) data stories that carry an audience through a pre-constructed narrative. (See Figure 5.8)

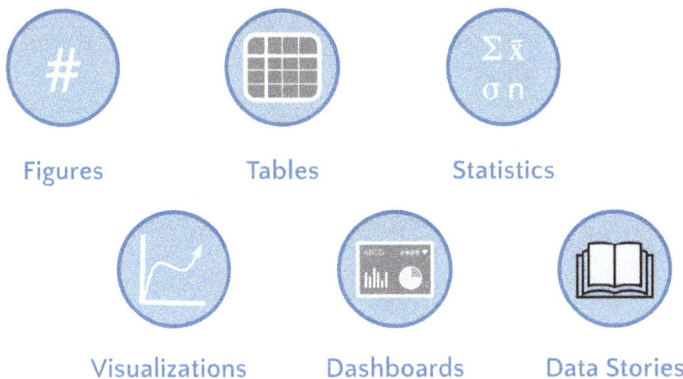

Figure 5.8. The Six Ways of Displaying Data

Each of these approaches has its own set of advantages and disadvantages. Sometimes, all your audience wants is a data point: one single fact or **figure**. This commonly arises when you're presenting to executive leadership. They don't want to see all the details, and bombarding them with elaborate visuals will only annoy them. They want to know about an overall sales goal, a bottom line, a completion timeline, a risk threshold, or a market condition. Just tell them that number, or at least start there and ask if they want you to go any further. Don't make the mistake of being an overly analytical engineer. Trust me, I know from experience.

Remember, though, we're all executives, just about different things in life. For example, you're the CEO of your own weekend afternoon. If you're trying to figure out whether to bring an umbrella on a stroll around the park with your friend starting at 2:00 p.m., you just need to know the chance of rain at that time in your town. You don't need a national weather dashboard—that would be too much information. It turns out that there's a 40% chance of rain. Better throw the umbrella in the back seat just in case.

Often, though, a single fact or figure begs the question: what about that same data point elsewhere, or at a different point in time, or in comparison to a different figure? For example, what is the chance of rain in your town at 1:00 p.m., or 3:00 p.m.? What about the chance of rain in your friend's hometown instead of yours? How about the temperature in both places?

This line of thinking leads to a list of figures or a grid with rows and columns: a **table**. For all the flak they get in the data visualization world, I think tables can be great in some situations. They capture and convey data with all its precision, and they connect us to the actual numbers. They enable us to do math in our heads to make a whole host of different comparisons. We can compute **statistics** from tables, such as measures of central tendency (mean, median, mode) and dispersion (range, standard deviation). We can compute percentages and percentiles and rates galore.

The downside of tables and statistics, though, is that they're not very good at revealing to us the underlying patterns or trends in our data. We simply cannot see the *shape* of the data with these forms of display alone. Furthermore, forcing us to do mental math in order to make comparisons can be a slow, cumbersome, and inaccurate process. Instead, we represent the data using graphical marks like points, lines, and areas. Then we use our data to encode these graphical marks using visual channels, such as position, length, size, angle, and color, among others.

What we end up with is a chart—a **visualization** of our data—which enables us to leverage the powerful visual cortex of our brain to quickly spot patterns, and to make very fast, relatively accurate comparisons. The data comes to life in front of us, in full, living color. It can be a beautiful thing, when done well. In the simple example of your afternoon walk, you look at a precipitation radar map of your county, and you see that the showers are concentrated to the east, on the other side of the hill running through the middle of your county.

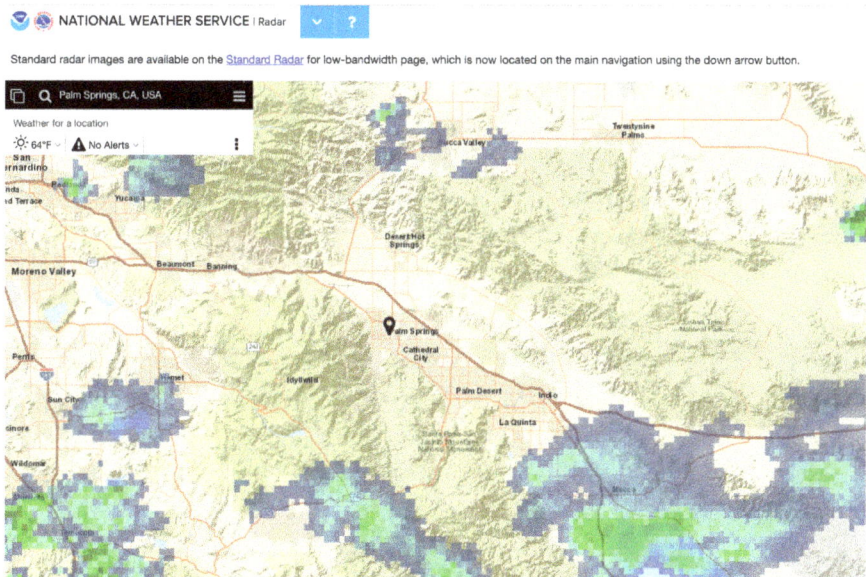

Figure 5.9. A NWS radar map of the Palm Springs area on March 30, 2023

In some situations, one single chart just doesn't reveal enough of the picture. Say, for example, that a map shows us how things differ by geography, but we need to know how things have changed or will change over time in order to make the decisions we need to make. A line chart would help us with this second question, but maybe we don't want to lose sight of the geographic element while we're comparing different time periods. So what we do is we look at more than one chart at a time. We place them side by side, called "faceting," or we superimpose them. Maybe we may also want to see facts, tables, and statistics along with these charts. We end up with a **dashboard**, and then we configure them to interact with each other: selecting (whether by mouse click or mobile tap) a county shape on the map filters the line chart to show change over time for that county only. Hovering over a point on the line changes the map to show the geographic distribution at that point in time.

Let's be honest, though: there are a lot of dashboards floating around out there, and they can be downright intimidating or confusing. There are situations in which we need someone to walk us through different aspects of the data step by step, as opposed to throwing everything at us all at once with a dashboard. In such situations we turn to **data storytelling**—the art of taking someone on a tour through the data. The presentation includes a set of findings sequenced together so as to form a cohesive narrative that gives the audience member a deeper understanding of the data. Data stories are very powerful when combined with the human element of the story. In our simple example, we watch a five-minute forecast on the local news, and we have a better idea of how the weather might change over the course of the afternoon.

We have considered the Six Ways of Displaying Data from the point of view of the audience member. Well, as a team leader, you need two things. First, you need to have a fairly high level of skill in communicating data yourself. Second, you need team members who also have this skill. With dedicated attention over time, along with guidance and feedback from others, you and your team can hone the craft of communicating with data. I guarantee you that it will be well worth your time and effort.

Foster a Mindset of Alertness about the Many Pitfalls in Data

Make sure that your team members are alert to the many inherent pitfalls involved with working with data.

To start this section, I'd like to ask that you indulge me as I take a trip down memory lane. I have spent a lot of time in my career helping people get started analyzing data. I began working in the business intelligence industry in January 2013, training journalists, bloggers, and students to visualize their data. By that point, I had already spent almost a decade in a continuous improvement role helping managers use data to reduce errors and waste in their processes. After my first book was published in 2014, I began teaching data visualization at the university level, helping most of my students develop a brand new skill.

What I observed over and over during that time is that people would make a lot of mistakes when working with data. That wasn't true only for the beginner analysts I worked with; it was also true for more experienced ones, and it was certainly true for me. The reality is that there are many opportunities for error, from the way the raw data is collected to the steps you take when you transform it to the many choices and details involved in the charts you make out of it to the way you think about and talk about your findings. These are just a few of the areas in which data pitfalls abound.

By the time 2015 rolled around, I was convinced that people all around the world were using powerful software programs to make massive mistakes with their data on a fairly regular basis. I began to be concerned that, in many cases, people wouldn't ever find out about the blunder they had made. They'd just go along their merry way, blissfully unaware of the misconceptions they had acquired.

Furthermore, I noticed that very knowledgeable professionals in the business intelligence and data visualization worlds were spending an inordinate amount of time and energy arguing about one single aspect of data analysis: chart type selection. Endless debates raged on Twitter about the merits of the connected scatterplot, or whether the pie chart should be banished forevermore, or in what situations the minimum value of a bar chart's y-axis could be something other than zero. A handful of other hot topics du jour surfaced and resurfaced.

I came to feel that these chart choices, while sometimes quite important, were just scratching at the surface of the real size and scope of the problem. In many cases, though, they didn't matter at all, and amounted to "rearranging deck chairs on the *Titanic.*" After all, if the data itself is fundamentally flawed, it doesn't matter where your y-axis starts. But these chart elements that are visible are all anyone in the industry was talking about. So in February 2015, I started writing my second book, *Avoiding Data Pitfalls*.

My father's health subsequently declined, and declined rapidly. He succumbed to glioblastoma at the end of April of that same year, and my life and my passion for work completely stalled. I'm grateful for my editor and publishers at John Wiley & Sons, because they stuck with me. It took a new relationship with my now-wife and co-founder Becky to jumpstart the writing process for me, and the book was published almost five years after I signed the contract to write it.

I'm proud of that book, because I made it across the finish line in spite of some considerable headwinds and setbacks in my life. And I'm proud of it because it has helped people come to terms with the fact that working with data isn't as simple and straightforward as we want it to be. Just because it's a tricky challenge, though, doesn't mean we should abandon it or give up hope. The idea is to forge ahead with an alertness, a willingness to acknowledge when errors have been committed, and a determination to fix it, learn, and grow.

You need people on your team who are willing to embrace a mindset of alertness when it comes to data. This attitude is so important, I included it in our short e-book and course *17 Key Traits of Data Literacy*. I have my mother to thank for the fact that we used the word "alert" in that program. I had originally chosen the word "wary," and it wasn't sitting well with me. It sounded too cynical and negative to me. She suggested a much more positive word, and we went with it.

So what do you need to do as a leader to foster a mindset of alertness to data pitfalls? Well, you need to make it okay for people to make mistakes, and you need to make it okay for them to admit it when they make mistakes. How else will your team build its immune system against data "infections"? If you establish a cutthroat culture, one that's harsh toward those who get it wrong, then your team won't really be alert to errors. Instead, they'll be wary of being caught making errors. That's a very different thing. You don't want your team members to be more concerned about being criticized than they are about fixing a mistake and missing an opportunity to learn.

You can model this for them. When you make a mistake, bring it to their attention. It doesn't have to be some long, drawn-out confession. Just say what happened, why it happened, and what you're doing about it. The idea isn't to tell them that it's no big deal, and who cares, anyway? The idea is to show them that getting it right is more important than being seen as right.

Get Your Team to Consider Data When Making Decisions

Encourage the people on your team to regularly advocate for the use of data in carrying out tasks and making decisions.

You've likely heard of the classic quote by Mark Twain about literacy: "The man who does not read has no advantage over the man who cannot read." In similar fashion, the team that does not use data has no advantage over the team that cannot use data. Do you and your team members use data when making decisions? Using data to make better decisions is what analytics is all about. That's the entire point.

I don't mean to get embroiled in a war of semantics, but I'm not a big fan of the phrase "data driven" in the context of decision-making. I much prefer the phrase "data-informed decision-making." The reason is that I don't believe data should be what *drives* our decisions. What should drive our decisions is our overall purpose, our mission, and the goals and objectives that stem from them.

Data can be like higher-octane fuel in the tank, though. Or, in a better analogy for the rest of the 21st century, it can be a supercharged battery that propels us faster and farther to our desired destination. Soon, it seems possible that data—as well as artificial intelligence powered by that data—will literally drive our vehicles for us. But for now, we humans are still in the driver's seat. That applies to both our vehicles and our businesses.

When people say that their organization needs to be "more data driven," what they mean is they want people within the organization to use data more often and more effectively when making decisions. They want to get better at moving from the Data level of the DIKW Pyramid to the Wisdom level. That's a very worthwhile goal. We should all seek to leverage data to the best of our ability when making decisions. But data is just one input of many in our decision-making processes. The experience and intuition of various team members are also inputs. So is time. And let's be honest, so is random luck.

The point in this section is to help you figure out how to get the people on your team to use data when they make decisions. We'll consider this from a process perspective in the next chapter. But for now, what are some ways you can encourage your team members to consider data when making decisions?

It starts with you yourself: are you leading by example? Are you seeking to use data when you make decisions, or are you solely relying on your gut? Now, it's important to acknowledge that there are situations in which your gut is all you can use: either you have no data, or you have no time, or you have neither. But those situations are relatively rare. You might not have as much time or data as you'd like to have, but you almost always have some of both.

As a leader, then, you model resourcefulness by making the most out of what you do have. You leverage data to the best of your ability in each given situation. If your team sees you doing this, then they'll want to follow suit. On the other hand, if your team sees you ignoring data, then they'll get the message loud and clear: we only want to *say* that we're data-informed; we don't want to actually *be* data-informed. You don't want them feeling that way.

Once you set the tone by your own behaviors, you then can set the expectation with your team members that they leverage data in their own decisions, too. But you will have to do your part. You can't just tell them to use data in their decisions while at the same time withholding access to the data and tools that they'll need to be able to do what you're asking them to do. That's a recipe for frustration and attrition. Your team members will leave your team if you set them up for failure in that way.

You also will want to avoid the mistake of enthusiastically encouraging your team members to make data-informed decisions when the reality is you haven't given them the authority to make decisions in the first place. It's important to consider, then, who you have granted the authority to make which kinds of decisions.

You may have heard of the **DACI decision-making framework**. This can be a useful tool when thinking about the people on your team and how they work together to make decisions. For each decision your team has to make, each person involved is given one of the four letters in the DACI acronym:

- "D" stands for **Driver**: This is the person on the team who is responsible for driving the decision.

- "A" stands for **Approver** or **Accountable**: This is the person on the team who is ultimately responsible for making the final decision.

- "C" stands for **Contributor** (or Consulted): These team members (there could be many of them) are responsible for providing inputs to the decision-making process, but they're not personally responsible for the final decision itself.

- "I" stands for **Informed**: These team members aren't part of the core team that will work together to decide, but the core team commits to updating them, because the decision might have an impact on them, too.

The main point is that as the team's leader, you need to know who's responsible for making which decisions. And then you need to make sure they get the message that data is a critical input to those decisions. You need to model that behavior in the decisions that you yourself own. And you need to make sure that your team members have the data and tools they need to do the same.

Give Your Team Plenty of Data Training Opportunities

Provide valuable training opportunities to help your team members develop the knowledge and skills necessary to effectively work with data in their roles.

Your team members want to be given opportunities to learn and grow. One way to provide them with such an opportunity is to give them access to great training offerings. While you may be concerned that a team member who receives training will want to leave, I recommend that you find a way to get past that fear. I agree wholeheartedly with the quote by British entrepreneur and tycoon Richard Branson in the epigraph of this chapter. The best strategy is to train your team members, and make it so that they don't want to leave.

What I've noticed, though, is that people in organizations often feel that they're not given such access to training. In fact, in our Data Literacy Score team-based assessment, out of 50 total scoring statements, the following statement has been given the second-lowest average score by thousands of respondents in the three years since we launched the assessment:

My organization provides valuable training opportunities to help me and my teammates develop the knowledge and skills necessary to effectively work with data in our roles.

That seems like an important gap to close, then. How would your team members score that statement on a scale of 0 (does not apply) to 10 (fully applies)? What are you doing, as your team's leader, to close that gap for your own team? Are you giving your team members access to great training opportunities? This isn't just something that would be nice to do. Some feel that it's core to what every organization should be doing for the employees who work there. Consider the following quote by Austrian-American management educator Peter Drucker:

Every enterprise is a learning and teaching institution. Training and development must be built into it on all levels—training and development that never stop.

As a data trainer, teacher, and instructor, I have seen the value that good training can provide. When designed and executed well, training can open participants' eyes to the art of the possible, and it can help them practice and hone that art. Unfortunately, training can also be a complete waste of time. We've all had to sit through that training session that was boring and didn't apply whatsoever to our job.

You'll need to learn to differentiate, then, between great training opportunities and those that are a waste of your team members' time, not to mention your budget. What are some characteristics of great training offerings? The following eight characteristics are hallmarks of well-designed training programs:

1. They have clear, relevant learning objectives.
2. They can be customized to your specific situation.
3. They engage trainees and solicit active participation.
4. They are delivered in a fun and engaging way.
5. They are taught by qualified trainers with deep credentials.
6. They include practical examples and real-world case studies.
7. They give trainees a chance to be assessed and receive feedback.
8. They feature follow-up touchpoints where learnings are reinforced.

When you find a training program that includes most or all of these traits, then go to bat for your team to get the budget to offer it to them. It won't be cheap. But, as Drucker also said elsewhere, "If you think training is expensive, try ignorance."

People Summary

A common refrain holds that an organization's people are its most valuable asset. That's also true when it comes to working with data. Do you have the right people on the proverbial bus? Are you giving them what they need, and then getting out of their way? All of these clichés completely apply to the challenge of being a great leader in the age of data.

I believe that you should empower and encourage the people on your team, but you should never blame or shame them. Chances are they're doing the best they can with what they have to work with, and considering other challenges that life is throwing in front of them. A little compassion and patience can go a long way. But so can a worthwhile challenge along with clarity of expectations. Try to get this balance right.

And when issues inevitably arise, seek to address them with a cool and calm demeanor that makes it clear that you have faith that everything will work out in the best interests of everyone involved. I can tell you that it will.

We considered seven different tactics you can implement to give and get the most from your team members. We can restate these seven tactics as seven powerful questions as follows:

1. What investments have you made to build your team's ability to work with data?

2. How are you helping them to become more fluent in the visual language of data?

3. Can you team members dive into raw data themselves, and if not, what are you doing about that?

4. How would you rate your team's data communication skills and what are you doing to increase them?

5. What are you doing to make sure your team is alert to the many data pitfalls that are out there?

6. How are you both modeling and enabling data-informed decision-making?

7. What high quality data training opportunities are you providing to your team?

Chapter 6

PROCESS

"If you always do what you did, you'll always get what you got."

–Marian Diamond

Leaders in the age of data think about what they do as a process, and they find ways to use data to improve their most critical processes. They apply this approach to their own processes as well as to those of their teams. How about you? Do you do that? It's an important question to ask yourself.

When things aren't going well, it's easy to sit back and complain about all of the obstacles in your way, or to blame others for a lack of progress. I've certainly taken that approach before. But that's a copout. True leaders take ownership of their situation and of the results they've been getting, and then they find a way to adjust their mindset and their actions in order to change the outcome. It's all about the process.

It's worth noting, however, that in the Data Literacy Score team-based assessment, out of the ten lowest-scoring statements to date, a full four of them are from the Process category. Statements relating to the flexibility of a team's processes, the documentation about those processes, and the way processes address deficiencies in data have ranked among the lowest-scoring statements in the entire assessment. Respondents seem to be more critical about those topics than about many others. I'm willing to bet that your team has issues with your processes as well.

In the epigraph of this chapter, we read the thoughts of path-breaking American neuroscientist Marian Diamond, who was the first to provide hard evidence that the human brain is capable of what we now call neuroplasticity—change, growth, and reorganization. If our brains are capable of such fundamental adjustments, then so are the processes invented by those brains. In this chapter, we'll consider powerful ways to do just that.

Design Recurring Processes that Leverage Data to Run Smoothly

Make sure that your team's recurring processes effectively leverage data in order to run smoothly.

Every team has at least a few recurring processes that they run on a consistent basis in order to achieve their goals and objectives and fulfill their mission and purpose. That process could occur on a yearly basis, an hourly basis, or anything in between. Here are some examples of processes that teams in different departments run on at various frequencies.

- A **yearly** performance review process run by an HR team
- A **quarterly** SEC filing process run by a finance team
- A **monthly** "month end" accounting process to close the books
- A **biweekly** payroll payment process run by a finance team
- A **weekly** systems backup run by an IT team
- A **daily** stand-up meeting held by a product development team
- An **hourly** product quality assessment by a manufacturing team

What processes does your team carry out over and over? Does your team participate in any cross-functional processes on a recurring basis? Do some of your processes recur, but not necessarily on a predictable or consistent basis? Perhaps your processes are triggered by an irregular event, such as is the case for spontaneous audits or employee exit interviews.

In each of these cases and many more, it might be helpful for you to review your processes, and to determine whether your team can use data to make them run more smoothly. Think of a smooth process as a well-oiled machine. It runs reliably, and its different parts and systems are all in place and operating properly. The oil in the machine ensures that the process runs without excess friction. Your data can be like the oil that lubricates the machine of your team's processes.

But what can data be used for, in the context of operation of recurring processes? Here are nine ways that data can be used to improve the way a recurring process runs:

1. Benchmarking against other processes
2. Establishing clear targets for process metrics and KPIs
3. Continuously monitoring performance and tracking KPIs
4. Improving the quality of manual steps in the process
5. Identifying anomalies, defects, and bottlenecks in the process
6. Collecting feedback on satisfaction levels
7. Automating some or all of the steps in the process
8. Simulating the process before it actually happens
9. Predicting how well the process will run in the future

These are just a handful of the ways that a process can benefit from the application of data in order to run more smoothly. Think of your team's processes, and think of ways to use data to improve them.

To give you a very simple example, my team sends out a newsletter to our email distribution list at the end of each month. The design and creation of each email is currently manual, and we use a customer relationship management (CRM) system to actually send the emails. We can use data in a whole host of ways to improve this monthly process.

First, we can use data to benchmark industry standards and set targets for our open rates and click rates. Doing so reveals that, according to MailChimp, an average click rate across all industries is 2.91%, and 2.90% for our Education and Training industry.[45] We can compare this with our most recent newsletter, which saw a click rate of 4.67%, and realize that our audience is highly engaged.

---

45. "Email Marketing Statistics & Benchmarks." Mailchimp. Accessed May 29, 2023. https://mailchimp.com/resources/email-marketing-benchmarks/.

For future newsletters, we can decide who receives the email based on their preference settings and their historical level of engagement. Do we send it to unengaged subscribers—those who haven't interacted with our emails recently? We can use data to find out what time of day and what kind of subject lines have corresponded with higher click rates historically. We can look closely at a heat map of the clicks for previous newsletters and identify patterns of interactivity.

These are just a few of the ways that my team can use data to improve this recurring marketing activity. The next step would be to automate these forms of analysis with A/B testing, and to make predictions about future performance based on decisions that seem to move the needle.

When you take the initiative to actually map out a process, you can see it from a bird's-eye view, and you can more easily identify places where data can improve the flow of that process. It doesn't have to be complicated. You just draw a box for each discrete step in the process, and you draw arrows connecting them in the order in which they take place. You can create a high-level process map, or a more detailed one. It isn't rocket science. But it will totally change the way you think about what you do. (See Figure 6.1.)

MONTHLY NEWSLETTER CAMPAIGN PROCESS

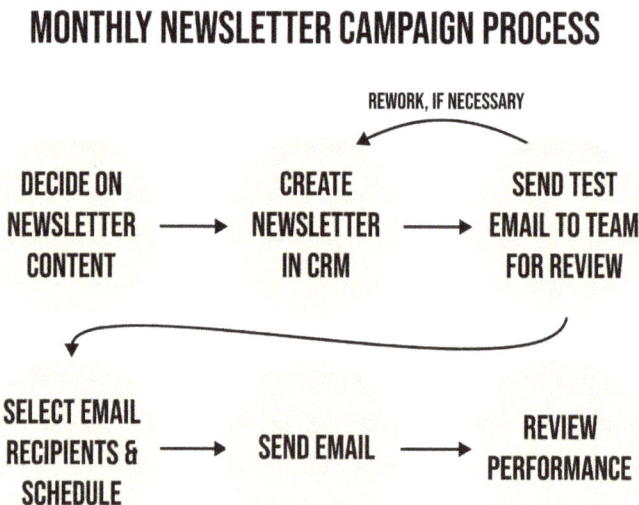

Figure 6.1. A simple process map of a monthly newsletter campaign process

You and your team will have processes that are as simple as this, and you'll have processes that are far more complex. The idea is to think about what you're doing as a process, map out that process, and then ask yourself how data can make it run smoothly.

Document the Way Your Team Works with Data

Make sure that the way your team works with data is well-documented and reproducible.

I've known a handful of people in my life who get really excited about creating thorough documentation. I can't say how much I appreciate these individuals. Because let's be honest, most of us don't get excited about creating really great documentation. The fact is, though, it's a huge relief when we're confused about something, and then we discover that, lo and behold, a handy guide already exists that walks us through it step by step.

Good documentation is a godsend for processes in general, but it's especially true of processes that involve data. The reason is that working with data is already very intimidating for many people. So is doing something they don't do very often. It's a "double whammy." Provide them with a how-to tutorial for a specific data task, and they'll feel much more comfortable tackling the challenge.

I know that's how I feel. One situation comes to mind. In the first three years after its launch, we ran the Data Literacy Score team-based assessment dozens of times. Each time, I was the one doing all of the data transformation and data analysis for the team. In an unusual fashion for me, I decided to create a detailed set of instructions for myself showing how to work with the data from the survey tool, complete with screenshots of each step of the process. When we evolved and matured the assessment the year after its launch, I took the time to update the instructions. That little guide has saved me countless hours.

The best part is that it didn't really take much incremental time to create it. I simply documented each step as I actually went through it one of the first times I ran through the analysis. I would estimate that it turned a two-hour task into a three-hour task on that one occasion. But I'm willing to bet that it has cut each successive round of analysis in half, turning that same two-hour task into a one-hour task each time. The best part is that I don't feel that dread that I know I'd otherwise feel leading up to the task. I just set aside the time, open the procedure, and take care of business.

Documentation is very helpful for data processes that you don't run through every week or every month. These less frequent processes can be particularly problematic because we don't get a chance to build a level of comfortable familiarity with them. We don't need a detailed procedure on how to brush our teeth, or how to login and check our email, because we do these things all the time, and they become second nature. But that's not the case with periodic data wrangling, or semi-recurring analytics, or quarterly or annual data clean up. Each time we do those things, we have to dust off the old workbench of our minds, and we have to get to work using tools that might be a little rusty.

There's another huge benefit to documenting the way you work with data: you can use those documents to train people on how to work with data effectively. Team members can quickly and easily cross-train on various data-related tasks, filling out your skill matrix in a way that adds flexibility and robustness to the situation. And when new people join your team, they can learn how to perform important data tasks with consistency. (See Figure 6.2.)

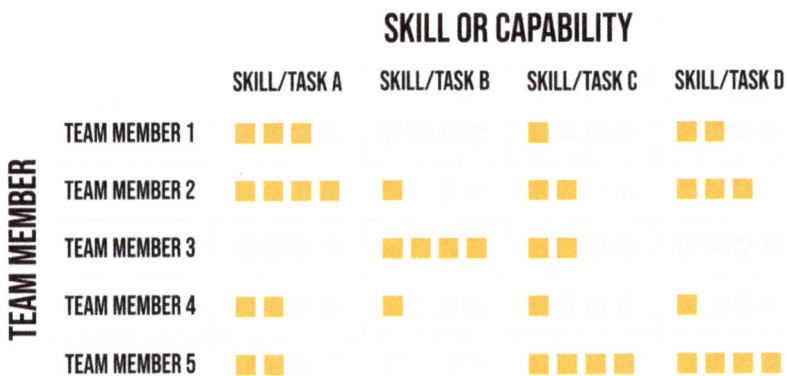

Figure 6.2. An example of a training or skills matrix

To repeat, very few people get excited about creating documentation, at least in my own experience. So when you find someone who is willing and eager to dive in and create helpful procedures and instructions for the rest of the people on the team, then take good care of that person. Having good documentation about the way your team works with data is something that they'll all come to appreciate. It does take a small investment of someone's time upfront. But you will all reap the benefits down the road. Your team will run through their repetitive data tasks more quickly, and they'll make fewer errors as they go along. It's a no-brainer.

Give Data a Prominent Voice in Your Team's Decisions

Consider what the data is saying, and make sure that anyone who brings relevant data findings to the conversation will be heard.

In the previous chapter, we considered the different levels of responsibility that people on your team might have when decisions need to be made. We applied the DACI decision-making framework to describe the roles of Driver, Approver, Contributors, and Informed. This is a helpful framework because it would be too simplistic to treat all team members the same when it comes to decisions. The reality is that they have different roles to play, and they have different levels of accountability relative to the decision and its outcome.

In this chapter, we'll apply that same logic to the decision-making process itself. This, too, can take on a very different type of flavor. Treating all decision-making processes the same would be overly simplistic as well. When you think about how well your team uses data when they make decisions, it can be helpful to think about three different levels of decisions: strategic decisions, tactical decisions, and operational decisions. (See Figure 6.3.)

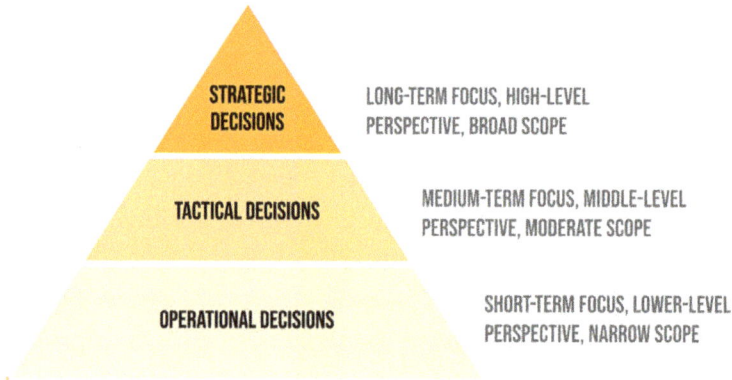

Figure 6.3. Three Levels of Decision-Making

Strategic Decisions

You may be involved in the shaping of your organization's overall aims and directions, and your own team has contributions to make to these strategic decisions as well. By definition, these decisions deal with your entire organization, and they trigger actions and follow-on decisions within each department.

When your organization seeks to make a strategic decision, the executive leaders are typically thinking about and affecting the course that the organization will take over the next two to five years. These decisions are very broad in scope, and they require those involved to take on a very high-level perspective.

Here are some examples of strategic decisions that organizations need to make and revisit on an annual basis:

- What are our goals and objectives for the next two to five years?
- Which customer groups will we seek to serve?
- Which markets will we enter, remain in, or leave?
- With what other organizations will we compete, partner, or merge?
- What's our competitive advantage and brand position within the marketplace?
- What's the structure of our organization's business units, departments, and teams?

- How will we make good use of capital and other resources to achieve our goals?
- How are we benefiting society and the environment?

Organizations commonly carry out an annual strategic planning process. This process is sponsored by the owners and the executive team, and it's often led by skilled and trained facilitators. Strategic planning is cross-functional in nature, and it involves each of the department heads and their chiefs of staff. Sometimes a dramatic shift in the environment, such as the onset of a worldwide pandemic or the spread of a disruptive technology, will trigger a round of strategic decision-making "off cycle," meaning sometime between the scheduled annual planning sessions.

But here's the question: do both your scheduled and your spontaneous strategic planning processes leverage data as a key input, or do they lack data as a key input? If your strategic planning processes haven't used data effectively in past years, how will you change them to better incorporate data in the next rounds or cycles? Who will identify data sources that can help you answer your most important strategic questions, and who will then gather these data sources together if they exist, or create them if they don't?

One unavoidable challenge with the questions involved in strategic decisions is that they involve a high degree of uncertainty and ambiguity. How will society and the marketplace shift and totally change in the coming years? What will other players in the marketplace do, and what won't they do? How will the needs and demands of your customers fundamentally change? What changes in availability and cost of resources will affect your financial viability?

There's no simple right or wrong answer to these questions. In similar fashion, the data sets you'll need to collect to help you think through these questions aren't always straightforward either. Making strategic decisions about the future involves forecasting and scenario planning. Your confidence in your strategic decisions and in the data used to make them may be quite low. But these decisions change everything. You can't claim to be a data-informed organization if your strategic decision-making processes ignore data.

Tactical Decisions

Even if you don't participate directly in your organization's strategic decision-making processes, you'll likely be involved in tactical decisions within your own department. Strategic decisions cascade down to the department and team levels, and trigger a whole host of tactical decisions that middle management must make in order to carry out the strategic direction set by top management.

Whether your organization choreographs this "flow down" of decisions in a deliberate and purposeful manner is another question, though. As nice as it would be to coordinate decision-making at various levels of the organization, the truth is that it's not always easy to foresee what tactical decisions will arise from the strategic decisions that are made. Furthermore, the trigger can be a delayed one: additional tactical decisions can often arise down the road.

For example, a strategic decision to enter a new market with an existing product will require the product and marketing teams to make many important tactical decisions to determine how, exactly, to accomplish that strategy. Some of those tactical decisions will be obvious from the onset. Others will surface later down the road.

- How will the company position the product against competitive offerings?
- What campaigns will the marketing team run?
- How will the sales team be compensated for expansion into the new market?
- Through what channels will the products be sold?
- How will the company deliver on their sales in the new market?

These questions are merely scratching the surface of the many tactical decisions that upper and middle managers will need to make in order to accomplish the strategic direction set by the executives. The success of the endeavor can hinge on any one of them. When you're involved in these tactical decisions, it can be helpful to stop and think of the decision as a process. What are the steps to the process? What are the inputs to each step? Which of those inputs are data inputs? Can data help us make the decisions we need to make as we carry out this process?

Operational Decisions

The next level down in the decision-making hierarchy involves day-to-day decisions that affect the routine functioning of the organization. These operational decisions are often made by employees who don't have a managerial role within the organization. Of the three types of decisions, these involve the shortest time horizon and the narrowest scope. But don't be fooled: they can be critical to your team's success.

As much as I dislike the term, I have to admit that some operational decisions are almost entirely data driven. For example, we seek to automate some of our more consistent, stable, and repeatable tasks and processes, and we can find ways to let the data dictate various parameters related to them.

Amazon, for example, does not have an army of workers deciding what books to recommend to you. Nor does Google turn to a massive crowd of employees to decide what ad to serve up to you when you search the internet. Those decisions are automated and optimized based on data, and they happen millions upon millions of times each day.

The role of the humans on the teams overseeing these data-driven processes is to use data to review their performance, to look for faulty or suboptimal outcomes, and to adjust the closed-loop system accordingly. Data drives the process itself, and it informs the review and continuous improvement of those processes as well.

Use Data to Continuously Fine-tune Your Processes

Apply data to find and eliminate wasteful errors, defects, and delays so that your processes improve over time.

I spent a substantial portion of my early career thinking about how to use data to continuously improve an organization's many and varied processes. This was a natural focus for me because I had started out as a mechanical design engineer, and in that role I had also used data extensively. Instead of using it to improve business processes, I had been using it to improve the performance of electromechanical parts and assemblies. So the transition from engineering to business process improvement was a relatively smooth one.

The story I like to tell is that my father, who paid for my undergraduate engineering school tuition, was more than a little surprised and confused when I told him that my job title at work was "master black belt." But it was true: I had joined the Lean Sigma movement, and I was working hard to eliminate wasteful errors and delays in manufacturing and office processes alike.

I learned a lot over that half a decade in my late 20s and early 30s. I learned about the Six Sigma methodology, and how to apply statistical approaches to solve business problems with data. I learned about the ideology known as **Lean**, developed in the middle of the 20th century by Taiichi Ohno, the father of the Toyota Production System. According to this approach, an organization identifies all the steps it takes to deliver value to its customers, and then it rigorously and ruthlessly eliminates any action that doesn't directly add value.

Non-value-added (NVA) steps and actions are known as *muda,* the Japanese word for waste. Traditionally, there are seven different forms of muda, and an eighth form has been added to the list more recently:

1. **Transport:** Moving products from one place to another, when such movement doesn't improve the products in any way, is wasteful and should be reduced or eliminated.

2. **Inventory:** Having excess raw materials and finished goods on the shelf does not add any value to the customer, and should be carefully managed.

3. **Motion:** Requiring employees to walk, reach, or lift too much in the production of a product does not add value, and might result in fatigue or injury.

4. **Waiting:** Causing available people or equipment to wait around for something to work on is wasteful, and should be minimized.

5. **Overproduction:** Producing more than is required or producing it much earlier than required leads to excess inventory and potentially obsolescence, and should be eliminated.

6. **Overprocessing:** Performing more work on a product than a customer would be willing to pay for is also wasteful and should be avoided.

7. **Defects:** Creating products that don't meet the quality requirements of the customer result in scrap and rework, and their root
   causes should be identified and eliminated.

8. **Underutilization of talent:** Failing to give an employee a chance
   to contribute according to their abilities and to improve and further develop those abilities is wasteful, and is an example of poor
   management.

You can tell that this list of wasteful actions was developed by people who spent much of their time in a manufacturing environment.
These forms of waste refer to raw materials, finished products, factory
equipment, and factory workers. And yet I found, time and time again,
that these forms of waste can be observed in an office environment as
well. Once your eyes are open to it, the similarities between the physical
world of manufacturing and the digital world of office work are easy to
identify.

- An overflowing shelf of work-in-progress (WIP) on the shop
  floor is similar to a growing queue of customer service tickets in
  a customer relationship management (CRM) system. Both are
  forms of excess inventory in a process.

- A product that fails inspection is like a loan application or web
  form with fields that contain typos and errors. Both are forms of
  defects produced by a process.

- Factory workers on a production line who are waiting for a
  malfunctioning machine to be fixed are analogous to a team of
  accountants waiting for the Wi-Fi in the building to come back
  online so they can finish closing the books for the month.

These are just a handful of connections that can be made. The keys to
unlocking the power of continuous improvement are 1) to see the work
being done as a process, 2) to identify what adds value in that process,
and then 3) to use data to reduce or eliminate anything that doesn't add
value. It's important to define value from the customer's point of view,
not the company or the worker's point of view.

Once you make the leap in your mind to a process-oriented way of thinking, you can marry that approach with an emphasis on data-informed decision-making to get the best of both worlds. (See Figure 6.4.)

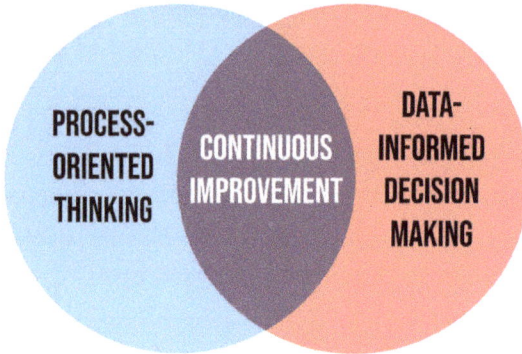

Figure 6.4. The secret to continuous improvement

What does it actually look like when these two ideologies come together? Once you and your team identify and "map out" your most important processes, you then collect data about the processes themselves to identify where there are forms of waste to be removed, and defects to be eliminated. There are many ways to use data to identify how you can improve a process. Here are just a few:

- You can carry out studies to find lead times and cycle times.
- You can conduct inventory counts of work involved in the process.
- You can sample work in progress to determine its level of quality.
- You can survey the people performing the process as well as the customers of the output in order to measure satisfaction levels.

One of data's best uses within an organization is to empower teams to continuously fine-tune their processes. Effective leaders know how to get their teams to see the possibilities of combining a process-oriented mindset with data-informed decision-making, and then to make that marriage a reality.

Put Processes in Place to Find and Fix Deficiencies in Data

Review and address deficiencies in your data and reporting so that your data becomes more and more valuable over time.

Next to your team members themselves, your data is your team's most valuable asset. But data is not perfect. It can be flawed in a whole host of ways, which we considered in detail in the Data chapter of this book. An ineffective leader throws up their hands and claims defeat. An effective leader finds a way forward.

Before using data in a recurring process or decision, it's important to evaluate its suitability for the task at hand. If you're under the impression that your data is perfect and flawless, then you likely need to look again, and you'll want to look closer. Once you realize that your data has defects and deficiencies, you have a tough decision to make: you have to decide whether or not it's good enough to use.

If it's not good enough to use, then you'll need to make improvements to it before you can use it. That might mean that this time around, you'll need to make do without it. On the other hand, hopefully you'll close the loop and make it better for next time. If it's good enough to use as is, then use it, but see if you can make incremental improvements to it so that it's even better next time.

These scenarios we have just considered can be represented by a simple flow chart, as shown in Figure 6.5. This flowchart is drawn to resemble a **closed-loop control system**, an important concept in control theory because it incorporates feedback in order to stabilize a process and improve its performance. It's easy to see why such a structure is powerful: the data that's used as an input is not only used, it is also improved as a result of the process, leading to better and better results.

HOW TO HANDLE DATA IN A RECURRING PROCESS OR DECISION

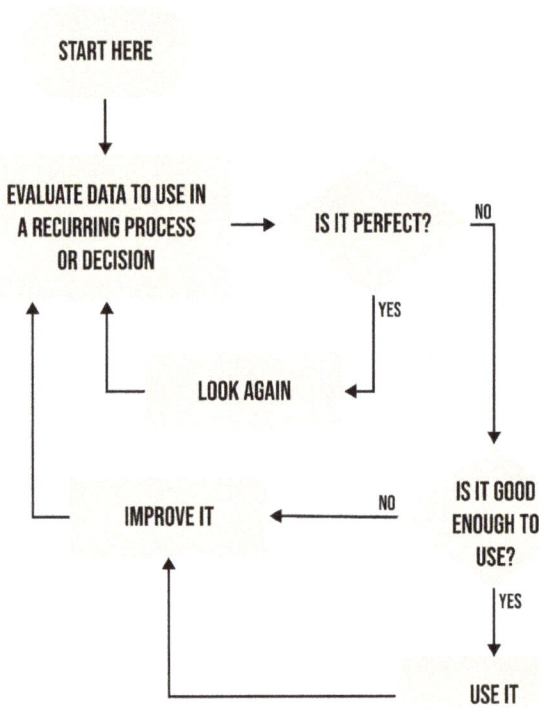

START HERE

EVALUATE DATA TO USE IN A RECURRING PROCESS OR DECISION → IS IT PERFECT? NO

YES

LOOK AGAIN

IS IT GOOD ENOUGH TO USE?

IMPROVE IT ← NO

YES

USE IT

Figure 6.5. A flowchart to decide how to handle data in a recurring process or decision

While your team may have the ability to implement the improvements to the data themselves, more often they do not, and you'll need to reach out to other teams and departments in order to "close the loop." This is where you can help your team identify who in the organization owns the data itself, the systems and processes that generate it, or the databases where it's stored. And then you can help the team by raising the issue with the leader who's in charge of that area.

If there are data sets that your team uses often, or that are critical for the successful completion of their most important recurring tasks, then you need to make sure that the quality of those data sets is continuously evaluated and improved. Are there automated tools that your organization can implement to identify and fix errors in the data? Are there ways to mistake-proof (or *poka yoke* in Japanese) the systems or the data entry steps that generate the data itself?

For example, consider researching and implementing the following approaches to make sure that the quality of your data is being actively managed:

- **Data cleaning** (also called data cleansing) is the process of identifying and fixing errors in data.
- **Data validation** rules can be implemented to make sure data conforms to certain requirements, such as "WA" for the U.S. state of Washington, not "Wa," "wa," or "Wash."
- **Data standardization** measures can be taken to transform data into a certain format, such as converting a date format like MM/DD/YY (for example, 08/30/19) to DD/MM/YYYY (30/08/2019).
- **Data deduplication** can be implemented to find and remove duplicate records in a data set.

These are all examples of actions that you can take that will result in improvements to the quality of your data. As helpful as they can be, you need to ask yourself what triggers them. Are they triggered by some event, like actually using the data, or do they happen at regular intervals, such as weekly or monthly? If there is no trigger, then they are "ad hoc" processes, and in my experience they probably won't happen. Or they may happen once and then never again until there is a big problem. A process without a trigger to initiate it can easily be forgotten.

Draw out the process steps, determine what triggers each one, and then assign responsible parties. That's the only way you can depend on these processes happening. That's data quality management. Taking an active approach to data quality management will pay dividends over time, as your data will continue to increase in both quality and value. Leaders who are able to advocate for these types of measures in a productive manner are priceless.

You definitely want to be this type of leader. You may have a few battles to fight now and then in order to make it happen. But if you spend time and energy building up good will and "leadership capital" within your organization, then you'll set the stage for some important data quality victories. Your team will thank you.

Grant and Revoke Access to Data Quickly and Accurately

Put in place processes that quickly and accurately adjust your team members' level of access to data and tools when they are hired or change roles.

One of the most frustrating experiences for someone working on a team is to be denied access to data they fundamentally need to use in order to do their job well. It can be very demotivating for them, and you can expect that part of their frustration will be directed toward you, their leader within the organization. After all, you're the one asking them to do something they can't do. It's your job to make sure they have what they need. If you can't get it for them, you have to change their job responsibilities to match what they can reasonably be expected to do.

Of course, there might be very good reasons why access hasn't been granted to them. If your organization has **data classification** processes in place, then certain data sets will be categorized as sensitive or private. Data privacy and security and good data governance policies and procedures restrict access to sensitive data sets in order to prevent costly breaches and, frankly, to avoid breaking any laws.

People should only be given access to sensitive data on a **need-to-know basis**, which means they need to have a legitimate business need in order to get access to it. If their business need isn't critical enough to match the level of sensitivity of the data they're requesting, then this needs to be communicated to them when their request is denied.

But so many times, people aren't given an adequate explanation when their requests are denied; they're just shown a brick wall. In cases like these, the principle of **least privilege** can be helpful: it might not be appropriate for them to get access to the entire data set, or to certain records or attributes, but other records and attributes might not be so sensitive. If so, perhaps partial access can be granted. It doesn't necessarily have to be treated in an all-or-nothing kind of way.

However, if the answer really is no, and there's no subset of the data that they can be given, then perhaps they can be given an anonymized or randomized version of the data. There aren't too many situations where such an approach would work for your team member, but cases do exist, such as for testing a newly designed graphic user interface, or GUI. The data used in the test doesn't have to be real; it just has to resemble the real thing.

If none of these alternate approaches work, then you're going to need to reassign the task to someone else who has the proper approvals and clearances. If there is no one else on your team who can get access to the data, then you'll need to talk to your own boss to get the associated task removed from your team's list of obligations.

Unfortunately, I've noticed that many times the lack of access isn't for any good reason at all. The data being withheld isn't sensitive at all, nor does it even contain any Personal Identifiable Information (PII). If such a situation isn't rectified speedily, you can expect your talented team member to leave your team. It's a maddening experience to have meaningless red tape preventing you from doing a good job. You want the kind of people on your team who aren't willing to put up with that. So the onus is on you to identify all the times when your All-Star players are being sent to the plate without a bat in their hand.

Of course, what's really needed is a balanced approach to data access. Team members who have both a legitimate need and the proper clearance for sensitive data should be able to get access to it quickly, with appropriate limitations on its use. When the employee's need for the sensitive data is over, access should be quickly removed. Data that has no PII and that's not sensitive in nature should be granted relatively quickly to those who need it. In many cases, data that's not sensitive in any way can be made available to all employees, completely removing the need to request access to it, and any associated delays.

Achieving this balanced approach requires a very nimble and efficient process. The cycle time of this process needs to be short, with no long queues or wait times. Since teams themselves are fluid, and since the needs of the business are fluid, data access needs to be fluid as well. And the accuracy of the process needs to be very high. Criteria around what amounts to legitimate use of sensitive data needs to be clearly spelled out so that there's no ambiguity about who should be able to access what. Since breaches can be very costly, there's little room for error.

If you work in a medium- or large-sized organization, chances are that you don't own the process to grant or revoke data access. You and your team members provide inputs to this process when you make a request, and you are the customer of the output. But you depend on others to actually carry out the process to its conclusion.

Additionally, more than one other group may be involved in carrying out the process. Perhaps HR owns a step or two, and so does IT. The handoffs between groups are often where the delays can pile up. **Segregation of duties** can be helpful from a security standpoint, reducing the risk that one person has excessive rights that allow them to tamper with the system. But it can also be challenging when it comes to process speed and efficiency, as multiple cooks in the kitchen can slow things down.

Your role, then, is to make sure to provide inputs that are as complete and accurate as possible, and to track the request as it moves through the process. It would help if you avoid the mistake of the boy who cried "Wolf!" How do you avoid that mistake? By asking for access to sensitive data only when it's absolutely necessary.

You also play a role in working with those who process the request, communicating the business need, and providing productive feedback to the process owner when issues arise. Now and then you might need to be the "squeaky wheel" for your team members to ensure things are moving along. But be careful how often you play that role, as people will learn to tune you out. Don't be the wheel that's perpetually squeaky.

You can help your team by building strong partnerships with leaders in HR and IT. Part of how you do that is by respecting and adhering to the data privacy and security policies that your IT department is responsible for overseeing. Doing so will only help you in those moments when you need to call in a favor and get emergency access granted for your team.

Make Processes Flexible Enough to Adjust to Changing Conditions

Design processes to be flexible enough to adjust along with the changing state of your data and technologies.

As we have seen, process-oriented thinking can be very helpful because it brings with it a level of rigor, consistency, and dependability. When processes have been established, fewer items on the to-do list fall through the cracks, and more items get done at a high level of quality. Instead of the "Wild West," the situation feels much more in control.

But process does have a dark side to it, consisting of red tape, bureaucracy, and inflexibility. Documenting every single thing that gets done with a formal and official departmental operating procedure can be quite stifling. There is a pendulum effect, and like so many other things in business and in life, you want to strike a balance between two extremes. You don't want everything to be chaotic and sloppy, but you also don't want it to be totally locked down and rigid.

One reason is that the environment itself isn't locked down and rigid. The climate and the situation is constantly in flux, and what worked well yesterday might not necessarily work well today. So you and your team will need to adjust to the changing conditions in which you're operating.

What are some ways you can do that?

Err on the Side of Simplicity over Complexity

You don't want to document every last detail of every minor task that your team does. You may think that this approach is good because it will result in consistency. But if you overdo it, you'll create a very rigid environment for your team members. When you document your processes, allow for some amount of flexibility in how things get done. Don't leave everything up to the whims of the team members carrying out the process, but don't be overly prescriptive about every step.

On my team, we prefer to make simple checklists for our recurring tasks, and only include detailed instructions and screenshots when a step needs to be done a certain way in order to get a desirable result.

Review and Update Your Policies and Procedures

As soon as you or one of your team members notices that the way something has been done no longer works, you have an event that should trigger a revision to one of your procedures. Maybe a tool has been added to your team's toolkit that makes a process much easier to perform. It would be a shame if your procedures continued to reference the old way of doing it. Or maybe a technology or system has been migrated or updated in some way. How frustrating it would be if one of your team members pulled up the instructions and they no longer applied.

All of these situations and many more should prompt you to make changes to your documented processes so that they're current and up to date. If you allow your procedures to become outdated, your team members will start ignoring them, and you'll soon find yourself back in the Wild West.

Train Your Team Members on Your Processes

There's no use having a bunch of perfect policies and procedures if nobody knows how to find them or how to follow them. Use a designated portion of your team meetings to give everyone a chance to brush up on the latest revisions to important team processes. Make sure everyone's on board with how things are changing, and why they're changing in the first place.

Training your team on your processes as they update also provides you with a good opportunity to find out how they feel about the changes that you're implementing. It would be even better to gauge team reaction before the change goes into effect, so consider reviewing proposed changes with the team beforehand.

Implement Agile Methodologies

In the 1990s, the software development industry was struggling with rigid and linear processes that weren't working as well for them as they had in the past. There were two trends happening at once that made their job increasingly difficult. First, what they were building was becoming increasingly complex. Second, the world was in a state of flux, as it is today, and the market demands would change before they were done. So they needed to come up with an approach that allowed them to work collaboratively, and make small adjustments as they went along.

A group of industry experts met at a ski resort in Utah in 2001 and developed the Agile Manifesto.[46] This document outlined a set of principles in which people would be valued over process and flexibility would be valued over adherence to a set plan. With Agile, the emphasis shifted to "early and continuous delivery of valuable software" that meets customer demands, even if the requirements change late in the process.

Agile was an answer to the handcuffs and blindfolds of overly cumbersome project management approaches, and it spread rapidly throughout software development in the early 2000s. Since then, Agile has been adopted by other types of teams as well, such as research and development (R&D) teams, design teams, marketing teams, and startup businesses. Agile doesn't have a set sequence of steps, but it typically involves short "sprints" of two to four weeks that allow teams to decide what they'll build, build it, test it, and then iterate. If you're finding that your own team's processes are too rigid, then something like Agile might be a better approach for your team.

Delegate Decision-Making as Much as Possible

One big mistake that team leaders make is that they hang on to all of the decision-making authority. This results in two problems. First, their senior team members get frustrated because they are limited in how they can contribute. Second, the team's work can grind to a halt whenever the leader is not available to make a final call. If every one of your decision-making processes has to go through you, then you are a bottleneck for your team. They won't appreciate it.

The answer is clear and obvious: find processes and decisions that you can delegate to others on your team. Use the DACI model again to shift responsibilities to others instead of hoarding it all for yourself. It's insecure and will cause you to be overworked and your team to be underutilized. You don't want either one of those situations. So trust your team. Allow them to make decisions. Do it gradually if you feel nervous about it, but get the process started. You'll be glad you did.

---

46. Beck, Kent, et. al. Manifesto for Agile Software Development. 2001, Accessed May 29, 2023. https://agilemanifesto.org/.

Put the Emphasis on Outcomes over Compliance

One of the main tenets of Agile is to put the focus on "working software over comprehensive documentation." Basically, the idea is that as long as your team is producing an end product that satisfies customers, it isn't so important that they perfectly follow a particular process while creating it. This also lines up with one of the main tenets of Lean, that the team should focus on doing activities that add value from the customer's perspective.

Yet it's important to acknowledge that there are some processes at work that do require strict compliance, such as those involving safety or security, or those related to legal or contractual obligations, or those mandated by a regulatory body. There isn't wiggle room for your team to do as they please in those situations. In a great many other situations, though, such strict compliance isn't necessary, and trying to enforce it will only cause frustration.

So it's important for the leader to distinguish and differentiate between processes that require strict compliance, and those that do not. If you allow for leeway and flexibility when rigidity is not necessary, then you'll be in a better position to ask your team to adhere rigorously when that's what's needed.

Measure and Monitor Process Performance

If a process is performing well, there's no need for it to be flexible. It just needs to keep doing what it's doing. As soon as it starts performing poorly, it needs to adjust. Ideally, it would anticipate the need to adjust, and do so before performance worsened. The problem with processes, however, is that they're often designed to operate efficiently in a very specific environment. As long as certain conditions in that environment persist, then everything is fine, and the process runs smoothly.

But, as I'm sure you've experienced, the environment has a funny way of changing, and then changing again. And so yesterday's processes aren't always capable of running smoothly in today's environment. How is it that we often find out that a process is performing poorly? Things just start falling apart. When that happens, there's no ignoring the problem anymore.

That's why a data-savvy leader keeps their finger on the pulse of the process. They closely monitor their most important processes, and they find out as soon as things start to go south. The smartest managers use **statistical process control**, or SPC, to keep track of their processes, because it allows them to know as soon as something is going wrong. Statistical process control is the application of statistical methods to track the performance of a process.

A common tool used in SPC is the **process behavior chart**, more commonly referred to as the **control chart**. With a control chart, various output parameters of a process are tracked on a special type of line chart that includes an "upper control limit," or UCL, and a "lower control limit," or LCL. These limits are not set by any person or group of people, but by the data values themselves. There are different ways to calculate the control limits, but I prefer the calculations developed by Walter Shewhart, as shown in Figure 6.6.[47]

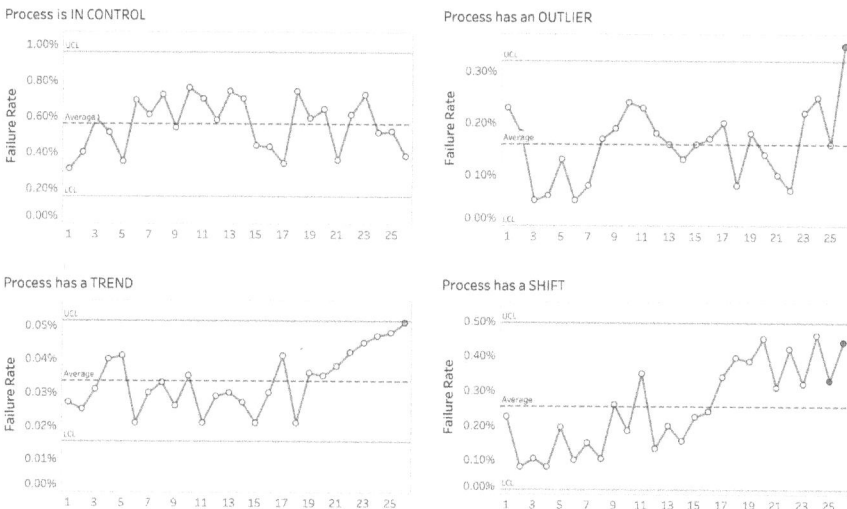

Figure 6.6. Process behavior charts (control charts) with various conditions

As you can see, a process can be in control, meaning there are no "signals" in the data, and all individual values lie between the control limits.

47. "Shewhart Individuals Control Chart." Wikipedia, last revised July 17, 2022. https://en.wikipedia.org/wiki/Shewhart_individuals_control_chart.

The variation we see in this chart is just routine variation, and nothing out of the ordinary has occurred. In the top right corner of Figure 6.6, we see a process that has an outlier, or a value that lies outside of the control limits. There is a very low chance that routine variation would produce such a value, so you can investigate to find out what might have caused this jump.

The bottom left chart reveals a trend, defined as six or more successive points that all increase (a rising trend), or all decrease (a falling trend) over the point that directly precedes them. Once again, it is not statistically likely that a process with nothing but routine variation would produce such an occurrence. And so you can investigate what might be causing the process to either rise or fall.

The chart in the bottom right hand corner of Figure 6.6 shows a process that has shifted. A shift is defined as nine or more points in a row that all lie either above or below the average line. This indicates that something has caused the process to shift higher or lower in the measurable output. Once again, you will want to look into what might have caused this shift. The chances that you'd see this if nothing special had happened is very low.

You can use this kind of approach to measure many different kinds of key performance indicators, or KPIs. You can use it to track output, such as units produced or completed. You can use it to track quality levels, such as failure rate or rework. And you can use it to track satisfaction levels, such as those obtained from a survey. The bottom line is that you want to watch the way your repetitive, recurring processes are performing, and you want to know as soon as something has changed, so that you and your team can react.

Process Summary

As a leader, you want to use the power of process-oriented thinking while avoiding some of the traps that can come along with seeing what you're doing as a process. Processes can bring dependability and reliability, but they can also bring rigidity and therefore fragility. Find ways to allow for continuous improvement and flexibility within your processes, and share the decision-making responsibilities with your team members, where appropriate.

Processes can leverage data, or they can ignore data. Processes that totally ignore data are just codified stupidity. Map out your most important processes, and then identify how data can make each step run better than it has up until now. This is the art of using data to improve processes. The reverse is also possible: you can use processes to improve your data. What processes do you have that identify and fix deficiencies in the data? What triggers those processes?

We considered seven different tactics to implement when leveraging the power of process as a leader. We can restate these seven tactics as seven powerful questions:

1. How do your recurring processes use data to run more smoothly?

2. To what degree have you documented the way your team works with data?

3. Does data have a prominent voice in your team's decisions?

4. In what ways are you using data to continuously improve your processes?

5. What triggers processes that improve the quality of your data?

6. Are your processes to grant and revoke data access both fast and accurate?

7. Can your team's processes adjust to the changing needs of the business?

Chapter 7

CULTURE

"Culture is what motivates and retains talented employees."
—Betty Thompson, CPO, Booz Allen Hamilton

---

believe I have saved the most important topic for last: culture. Perhaps only ethics can compete with culture in importance. When it comes to your leadership style and approach, ask yourself what kind of culture you want to create, and then think carefully about how you want to create that culture.

It's important to remember that you're not creating your team's culture in a vacuum. Perhaps your team is only part of a larger team or organization. In that case, your team's culture will be heavily influenced by the organization's overall culture. You have to find a way to work with that. And even if you are the CEO, and your team is the entire organization, you will be creating a culture that fits within an industry, within a society, and within the world itself. All of these broader cultures will tug and poke and push your own team's culture. It's up to you to harness some of those cultural forces and repel or counter others. It's not easy, and it's not for the faint of heart. I wish you the best of luck!

When it comes to data, if you want to be a successful leader in the era of data that we live in, you will need to find a way to make data your team's friend rather than its foe. Data can supercharge your team's culture, and data can sabotage your team's culture. Which will it be? In order to leverage data in your team's culture, you'll need to carefully consider the following seven factors.

Provide Access to Thriving Data Communities

Give your team members a chance to connect with others around data and its value as a resource.

Community is an interesting word to me. When I was a child in the 1980s, I used that word almost exclusively to refer to people who lived in the same area. At school we were required to do community service to successfully pass our ninth-grade health class. After school we would go to the community center to play basketball or tennis. In the evenings we would take extra classes at the community college. All of these uses of the word involved people who were in the same physical place at the same point in time. It was only because of that co-location that those people formed a community.

In today's digitally connected world, *community* has come to represent much more, but also in some ways much less. Our technology has greatly expanded the application of the word because it's now possible for people to be anywhere in the world and still be part of a community. At the same time, at least in my opinion, it has stripped the word of some of its potency, as people interacting online don't necessarily experience the same richness of interaction as they do face to face, nor can they turn to each other as neighbors for a stick of butter or a helping hand moving the couch.

We have seen a fascinating trend in recent years: technology companies seeking to harness the perceived power of community in order to foster enthusiasm among their user base and thereby to grow their sales. A thriving community of users is seen as a huge competitive advantage for companies. "Community-driven companies will always win," says Chris Anderson, former editor in chief of *Wired* magazine.[48]

---

48. "Chris Anderson on Why Community-Driven Companies Will Always Win." BCG Global, July 9, 2014. https://www.bcg.com/publications/2014/technology-strategy-innovation-chris-anderson-why-community-driven-companies-will-always-win.

The data world is no stranger to this trend. It's presently full of software companies who are striving to be "community-driven," and who eagerly hand out trophies, badges, honorary titles, and company swag to those who loudly trumpet their products on social media. I must admit that I'm slightly jaded about all of the hoopla, and the way users have been marshaled into service as unpaid company spokespersons. There's no doubt that many of these community members benefit in kind from the participation and the accolades. But the built-in *quid pro quo* can cast a long shadow over the entire party.

The good news is that there are ways to ditch the seedy underbelly of tech communities of practice, while retaining the helpful aspects and benefits that they bring. First, you want to get rid of any expectation of reward or promotion tied to mere participation in the community. The goal of the community is not the community itself, but rather the growth and connection of its members. If those members get promoted, it will be because they applied what they learned in impactful ways, not because they attended community events and checked all of the membership boxes.

Setting all that aside, we can see why communities—even online ones where the members never shake hands—can be powerful. They foster another, related word: communication. No one wants to work in an organization that's full of isolated silos where people don't communicate. The "work silo" is a byword in the business lexicon, evoking images of tired office workers stuck in drab cubicles and unable to connect with others outside of their small, increasingly inbred team. In this environment, ideas stagnate, opportunity for growth evaporates, and people feel disconnected. It's a lonely, boring culture.

Instead, find ways to connect people across the company who can learn from each other. Here are a handful of ways you can get started:

- **Have them learn from the experts:** Set up a training session or a hands-on workshop with a professional trainer, invite a guest speaker to share thoughts with the group related to their area of expertise, or start up a data book club to go through a book together.

- **Have them learn from each other:** "Skill builders," "skill pills," "lunch-and-learns," "brown bag sessions." I have heard them called these names and many more. Whatever you call them, have your team members teach others when they learn how to do something useful, and give them the chance to learn from others as well.

- **Have them help each other with their work:** Start an informal data mentorship program or set up internal discussion boards where they can ask each other for help. It can be a relief to be able to tap into an internal base of knowledge and experience when faced with a daunting data challenge.

- **Have them participate in contests and hackathons:** Give them a weekly challenge to complete, or facilitate a competition for individuals or teams to go head-to-head and apply their skills. Be sure to make it fun, definitely make it optional, and if you can, make it relevant to a real business need.

These are a handful of ways you can give your team members a chance to connect with others across the organization around data and its value as a key resource. All of these initiatives take time to plan and execute, so work with your peers to identify individuals on your respective teams who can play an active role as community builder.

Connect Your Team's Identity and Vision to Data

Make data a core element of your team's identity as well as its vision of the future.

A large part of a culture stems from the creation of a shared identity and vision. When we come together, who are we, collectively, and what do we have in common? What vision do we share about the future, and what do we all want to see happen?

Of course, we can find answers to these questions that are obvious and uninspiring. Who are we, collectively, and what do we have in common? Well, together we form a team, and therefore by definition we're all employees who work on that team within an organization. What do we all want to happen? We all want to succeed and grow—individually and, for some of us, collectively as well. Ho hum. Does work really have to be so motivational?

In our society today, yes absolutely, that's what many of us have come to expect from our careers and from our employers. In our broader culture, at least in the country where I live, we tend to want our job to be fulfilling, and we want to believe that our place of employment is making a positive difference in the world. When we wake up in the morning, we want to feel like we'll spend our day doing something worthwhile, and that we'll be teamed up with others who are excellent in various ways.

That's a high bar, if you think about it. We're asking a lot of our careers and our employers. These high expectations can lead to disillusionment as differences between us arise, and when inevitable challenges set in. How do you deal with that reality, as a leader?

One way is to create a positive, supportive culture, but not to overdo it. You can't demand continuous enthusiasm from your team members, and you have to allow them to have bad days too. So there needs to be understanding and balance. Some days, your team members are going to love their jobs. Other days, not so much. That's okay.

This very much applies to the overlap of culture and data. What does that overlap look like? Part of your team culture—not all of it, by any means—should embrace the power and value of data, and it should paint a picture of a future where data is used more and more effectively as time goes along. Why should your culture include these aspects? Because data is increasing in quantity and quality, tools are increasing in capabilities and availability, and none of this is going away, barring some catastrophic event. But your data can be a disaster. Some days, your team members are going to love their data. Other days, not so much. That's also okay.

Your team members are practically begging for an opportunity to get past their "dataphobia" and to learn how to harness the power of data in their careers. As their team leader, you play a huge role in making that happen. Data is messy and it's scary, so there will be plenty of disillusionment with this aspect of your culture, just like the other aspects. Build into your culture and your values an acknowledgment of the difficulties of data, and a willingness to tackle it anyway. Don't talk about data like it's some kind of panacea.

If you stop and think about it, the fact that your team members are all on the same team in the first place is actually quite remarkable. There are eight billion people on the planet. What are the odds that those people all ended up together? You can say that it had to be *someone* who joined the team, but in a sense it's pretty amazing that it turned out to be *them*. So celebrate this moment in your lives when you get to be part of the same team culture. And find a way to ingrain data in that culture, in spite of its unruliness.

Recognize and Reward Data Successes

Value and reward those who are able to make effective use of data to help you achieve success.

It's one thing to say that you want the people on your team to use data effectively. It's another thing to actually reward them when they do. Of the two forms of influence, telling them what to do is much less effective than rewarding them for doing it. What's more, if you only do the former and not the latter, then you'll almost surely reap the harvest of a resentful workforce.

Asking for something from those who work for you and then failing to reward them when they give it to you is like ordering a meal at a restaurant but not tipping your waiter, or even paying for the meal at all. Who would do that? Similar to the case of a patron ordering a meal, there's an exchange happening here: you give your team members responsibilities and direction, they put their time and effort into their activities, and then you make sure it's worth their while. You have to do your part.

Now you might say, "Why should I reward them for doing their job? Shouldn't their salary be enough of a reward?" There's some truth to that sentiment, and I'm not advocating for throwing cash at people every time they perform adequately. But I believe such a mindset can be very limiting. Let me explain.

The fact is that everyone responds to a mixture of motivating forces. That's true for your team members, and it's true for you and me, too. Each of us responds to these motivating forces differently, though. Some of these motivators can be oriented toward avoiding negative outcomes, and some of them can be oriented toward obtaining positive ones.

Positive outcomes often involve financial benefits, but not always. There are a wide range of positive motivators that you can use to incentivize your team members to do a great job, and to reward them when they do. We've identified eight of them here. Let's consider them one by one.

8 FORMS OF REWARDS TO MOTIVATE YOUR TEAM

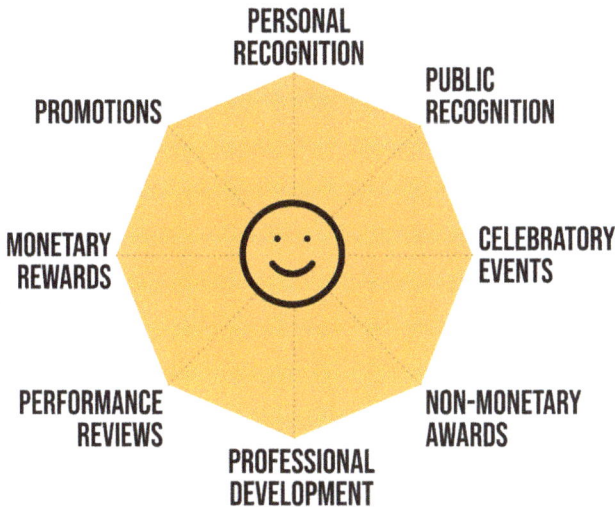

Figure 7.1. 8 Forms of Rewards to Motivate Your Team

1. Personal Recognition

When you acknowledge verbally to someone on your team that they have done a great job with data, you reinforce that behavior for that one person. There are many ways to do this. You can express appreciation for a job well done in an email to them, in a one-on-one meeting with them, or in the hallway in passing. This type of reward alone can be enough to recharge that individual. It communicates to them that their efforts are noticed, and that their contributions are appreciated.

Praise, whether delivered verbally or in writing, is more meaningful to some people than to others, and it can hit or miss depending on the situation. So some care needs to be given when using this form of reward, especially when it's all that you plan to give. Have you given them a lot of praise recently? Have they been at their current salary level for a while now? If so, your praise, as sincere as it may be, can fall flat.

2. Public Recognition

If you want to reinforce effective data working behavior for the whole team, instead of just one individual, then you can call out their achievement publicly. You can do this in an email to the whole team, in a team meeting, or in any situation in which the whole group is together. This adds a totally different element to the reward: recognition in front of their peers.

Being distinguished from the rest, even momentarily, meets a different social need than having your boss recognize your efforts. We all wonder whether we're as good as everyone else from time to time. So it can be very gratifying to get an affirmative answer to that question now and then. As the oft quoted saying goes, though, "Comparison is the thief of joy." So once again, you want to be careful how you go about setting individuals apart from the rest. Healthy competition on the team can be a good thing, but it can quickly descend into harmful pettiness if you pit one person against another. When pumping someone up, never put anyone else down in the process.

3. Celebratory Events

A way to put an exclamation point on public recognition is to hold team lunches or parties to commemorate the win. Maybe the win was achieved by one person alone, but more than likely there are multiple people involved. Either way, you want to make it clear that every win is a team win. Rewarding the whole team can be a great way to make that message loud and clear.

It's important, though, to call out the specific people who contributed to the win. Everyone celebrates, but not everyone gets congratulated for a job well done. After all, if everyone benefits equally every time, then some team members may decide to sit back and ride the coattails of others. Those doing all the heavy lifting will get tired of carrying the load for everyone else. These are delicate balances to achieve. You would likely treat a team that's full of high performers differently than you'd treat a team with just one or two.

4. Nonmonetary Rewards

When your team members achieve great successes with data, you can always reward them by giving them awards, prizes, or other gifts. Your organization might have some form of "all-star" accolade, badge, or trophy that you can give out, similar to the "employee of the month" placards you see posted on the walls of some establishments. Some people are very motivated by these awards, while others are not.

5. Professional Development

Professional development opportunities can be a very useful type of nonmonetary reward. If you provide high achievers with learning opportunities, especially when completion involves a meaningful form of credential or certification, then you can really propel their career forward.

Give high achievers the green light to sign up for a great online course on the company's dime, send them away to a university for an executive program, or work to gain approval for tuition reimbursement for their undergraduate or master's degree. Even a simple shipment of the best books in their discipline can be a wonderful form of reward that pays dividends.

Personally, I have received this form of reward at various points in my career, and I wouldn't be where I am today if those opportunities hadn't been presented to me. Just make sure your team member has enough time to actually take advantage of these opportunities. It can be incredibly demotivating to have a professional development opportunity dangled in front of you, while being so overloaded with work that you can't possibly use it.

6. Performance Reviews

It may sound obvious, but when someone on your team performs at a high level, you should give them a high rating in their annual performance review. It's not always so simple or straightforward, though. Perhaps they performed at a high level once in the year, but not for the rest of the year. In that case, their big win should be called out in their review, but you'll need to explain why other deficiencies affected the overall score.

Additionally, you might not be using a clear set of objective criteria to gauge their performance. Such a dearth of performance metrics means the scores are highly subjective. That's a much more challenging scenario for you as the manager, and a good reason to avoid such fuzzy situations in the first place. In general, though, when a team member has been doing great work with data, it makes perfect sense to call that out on their performance review, and let them know that these efforts have translated into favorable ratings.

The last two forms of recognition that remain are the ones that people typically hope for the most. They're also the ones that inexperienced managers assume are all they have to give. By now, hopefully I've reminded you that there are other forms of rewards that you can give out when people on your team achieve big wins with data. As nice as those nonmonetary forms of recognition can be, though, at the end of the day you want to reward your highest performers with either a raise, a bonus, or another form of cash award.

7. Monetary Rewards

There are a variety of types of monetary rewards you can give to your high performers. You can give them a raise (an increase in their salary), give them a one-time bonus payment, or award them stock or stock options. These are the main forms of monetary rewards you can give, but there are others.

In organizations that have a tightly controlled annual performance review process that allocates a fixed amount of funds to divide among all team members, the manager is presented with a challenging "zero sum game" situation: giving a higher raise to one person on the team means that there's less to go around for everyone else. This involves a delicate balancing act that isn't always easy to maintain.

For this reason, you might want to fight for an "off cycle" raise for those whose current pay is far below what their data contributions would warrant. When an individual team member's raise is disconnected from everyone else's midcycle, you don't have to "rob Peter to pay Paul," as the old saying goes. You'll likely still have to work hard to justify the midyear bump in pay to your head of finance. The fact that you have to go to bat for your team member only adds to the value of the reward, though. If it were so easy to secure a raise, an underpaid employee would be justified in wondering why you hadn't already done so for them.

A cash bonus can be another way to increase their compensation in a particular year. Since a bonus is a one-time payment, the employee's salary isn't affected. For this reason, bonuses might not hit the same exact payroll line item account on the company's ledger, and therefore might be subject to less scrutiny than payroll itself. This depends, of course, on your organization's accounting practices and current financial strategy regarding the management of overall compensation expenses. So you'll have to "feel that out" by talking to your own boss as well as midlevel managers in your finance department.

If you're getting nowhere with both raises and bonuses, you can con-sider using funds from your own team's budget to give out gift cards to your high-performing employees. Just make sure to check with human resources and finance to see what dollar amounts would result in tax implications or other complications.

Securing monetary rewards for your team members can be hard work. They won't always be aware of that difficulty, nor can you expect adula-tion or even appreciation for your efforts on their behalf. As humans, we have a way of feeling we deserve whatever we get, and our expectations about what is fair compensation can adjust very quickly. My personal approach as a manager securing a raise or bonus for my direct report is to assume that they'll feel that whatever they get is long overdue, and not quite enough. That way I'll be pleasantly surprised if they're thrilled and appreciative. I've just learned the hard way that this reaction is surpris-ingly rare, and that any appreciation they do feel will be short-lived as they settle into their "new normal."

8. Promotions

The last form of reward we'll consider is the most significant in terms of career development. When someone on your team consistently over-achieves with data, it might be time to move them higher in both title and responsibility. They typically have to "repeat the feat" at least once to warrant a promotion, though. If you use this form of recognition to reward them after a single accomplishment, then make sure that their one big success signals unequivocally that they'll now be able to take on a higher level of responsibility on the team.

I don't subscribe to the philosophy that a person should have to take on the responsibilities of a higher job before actually being given the title. In practice, this is what often happens in larger organizations. This could be due to a stingy boss, of course. But it could also be due to cost controls that require even a generous boss to build a veritable dossier to justify any promotion. It's true that these financial controls can serve as effective checks and balances against managers who are too eager to please their team members ("You get a promotion! You get a promotion! And *you* get a promotion!"). But they can also result in frustration and attrition of the very people you want to retain more than all the rest. Once again, your job as a data-savvy leader is to gather the evidence you need to promote your team members who deserve to be promoted.

We have considered all of these forms of recognition in the context of rewarding those who achieve great successes with data, but they can apply to high performers of all types. One additional factor to consider is that you might want to use them when the right behavior—effective use of data—doesn't necessarily result in a big win, for one reason or another. After all, it's the behavior you want to encourage and incentivize, not just the outcome.

Either way, you send a strong message to your team that you and the organization value their contributions even more when they take a data-informed path. Actually rewarding such behavior embeds that preference in the very culture of the organization. Failing to reward it signals that you're not really that serious about data, and the behaviors that do get rewarded are the ones that will continue shaping the culture.

Sponsor Data Learning and Development Opportunities

Commit to fostering a climate of continuous learning and development with respect to data.

As mentioned in the previous section of this chapter, giving your team members plenty of opportunities to develop their data knowledge and skills is a great way to reward them and thereby foster a data culture within your organization. As mentioned in the People chapter, access to data training opportunities seems to be lacking in many organizations, as evidenced by the fact that this is the second-lowest-scoring statement to date out of the 50 statements included in our Data Literacy Score team-based assessment:

> *"My organization provides valuable training opportunities to help me and my teammates develop the knowledge and skills necessary to effectively work with data in our roles."*

Since we've already spent some time thinking about how *training* can help your team members grow, in this section we'll consider the challenge from a slightly different angle: what aspects of your team's *culture* foster an openness to learning and development? I can think of three principles or "norms" that you should seek to cultivate in order to send the signal that learning is valuable.

1. **It's good to be curious:** First and foremost, you want your team members to know that you value curiosity and a willingness to learn new things. The whole point of using data is to learn something new. If people aren't curious, then they won't value data as much as they should. You can screen for this attitude in your hiring process by asking for examples of times when a candidate discovered something unexpected and exciting. Next, give them a chance to ask you questions about your team and the work you're doing. How would you rate the quality of the questions they ask? Now apply this same approach to your team meetings.

2. **It's okay not to know the answer:** Hand-in-hand with the previous principle, your team members need to know that it's perfectly understandable that they don't already know the answer to every question that might come up. Not knowing the answer is not necessarily a problem. If they feel that it's never acceptable to say, "I don't know," then you definitely have a problem. You have to model this mindset for them. Some leaders and managers never want to show a chink in their armor, so they avoid ever giving the impression that they don't know something, no matter how small or insignificant the detail might be. If you project such a know-it-all persona, you can expect them to mirror that right back to you. The problem is that nobody likes a know-it-all. An even bigger problem is that a whole team of know-it-alls is going nowhere.

3. **It's about getting it right, not being right:** An unchecked ego can be very damaging to a team's morale and to its ability to collaborate, and it's even worse when it's the leader whose ego is run amok. If you as the leader set the tone that it's more important to get it right than to be right, your team will feel more comfortable admitting when they, too, have encountered a valuable learning opportunity. On the other hand, if you need everyone on the team to think that you're right all the time, they'll become guarded about the times when they themselves don't have it quite right yet. Imagine how many valuable learning opportunities this fear-based environment will cause your team members to miss out on.

I believe that when it comes to data, it's much more efficient to learn by doing, rather than by reading or listening. That's not to say that reading about data skills and listening to others talk about them can't ever be helpful. It's just that in order for it really to stick, people need to try it for themselves.

I wonder whether the ancient model of **apprenticeship** is a valid model for learning how to use data effectively. According to this model, an inexperienced apprentice is paired with an experienced artisan mentor who "shows them the ropes." The focus is on deep observation of the mentor followed by guided, hands-on practice by the apprentice. The cycle is complete when the mentor provides feedback on the work done by the apprentice. The overall process involves many such cycles until the apprentice becomes the artisan, and eventually the mentor of their own apprentice.

Such a process can take a lot of time, depending on the particular craft the apprentice is trying to hone, as well as the desired level of that craft. But as long as the mentor has sufficient expertise and patience, apprenticeship can reduce much of the risk associated with superficial and inadequate learning that would otherwise lead to shoddy work.

It will be interesting to see how quickly AI assistants such as large language model (LLM) chatbots like ChatGPT will revolutionize the learning process when it comes to data skills the same way AI "co-pilots" have revolutionized computer programming. There's much promise to this approach, but also many problems and pitfalls. I outlined what I feel are a number of both the pros and the cons in my book and course *ChatGPT Basics.* In a nutshell, the use of AI raises concerns related to data privacy, copyright infringement, and inclusion of societal biases and inaccuracies, among other issues.

Nonetheless, being able to ask an AI tutor for advice and tips for working with data can greatly reduce the amount of time it takes to learn. Consider the following two prompts entered into the GPT-4 model of OpenAI's ChatGPT. In the first prompt, I ask ChatGPT to provide five principles to keep in mind when choosing what type of chart to use to present my data to my audience. I have specialized in data visualization for over a decade, and I can attest that the principles it generated provide a pretty good starting point.

Model: GPT-4

What principles of data visualization should I keep in mind when choosing a chart to represent my data? Give me the top 5 rules that will steer me toward an effective and impactful chart type and away from a poor choice that confuses my audience.

1. Understand your data and goals: The first step in choosing an effective chart type is to understand the nature of your data (e.g., categorical, continuous, time-series, hierarchical) and your goals for the visualization (e.g., comparison, distribution, correlation, trend analysis). This will help you select a chart type that best represents the patterns and relationships in your data.

2. Keep it simple: Opt for simplicity over complexity, selecting the most straightforward chart type that effectively communicates the desired message. Avoid unnecessary visual elements, 3D effects, and overly complicated designs that may confuse your audience or obscure the information you're trying to convey.

3. Prioritize readability and clarity: Make sure your chosen chart type is easy to read and understand, with a clear visual hierarchy, appropriate labeling, and a coherent color scheme. Choose chart types that make it easy to compare data points, discern patterns, or identify trends, such as bar charts, line charts, or scatter plots.

4. Consider audience familiarity: Be aware of your audience's familiarity with different chart types, and choose one that is likely to be easily understood by them. For instance, while you might be comfortable using a more complex chart type like a Sankey diagram or a radar chart, these may not be as accessible to a general audience as a simple bar or line chart.

5. Ensure accurate representation: Make sure your chosen chart type accurately represents the data and doesn't lead to misinterpretation. For example, avoid using pie charts for comparing data across categories, as it's difficult to judge the relative sizes of pie slices accurately. Similarly, be cautious when using area charts, as they may inadvertently suggest a greater emphasis on the magnitude of change between data points.

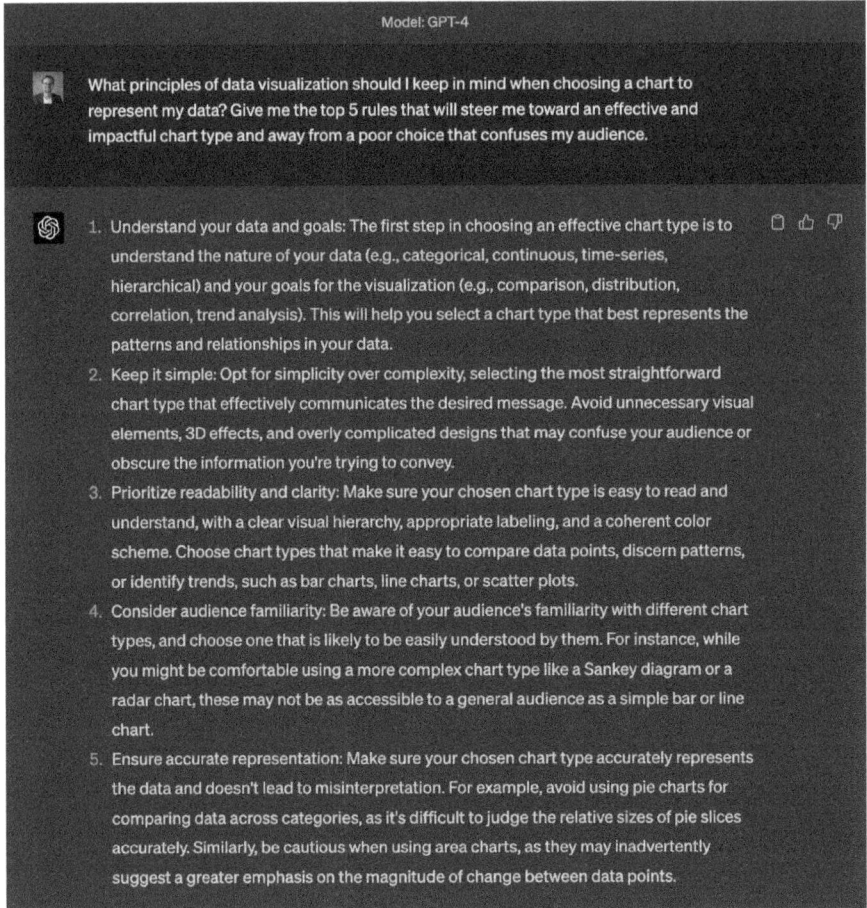

Figure 7.1. Using ChatGPT for advice on visualizing data

In the next example, I begin a multi-round set of exercises to learn SQL (Structured Query Language). These new "AI tutors" can go a long way to helping us develop our data skills.

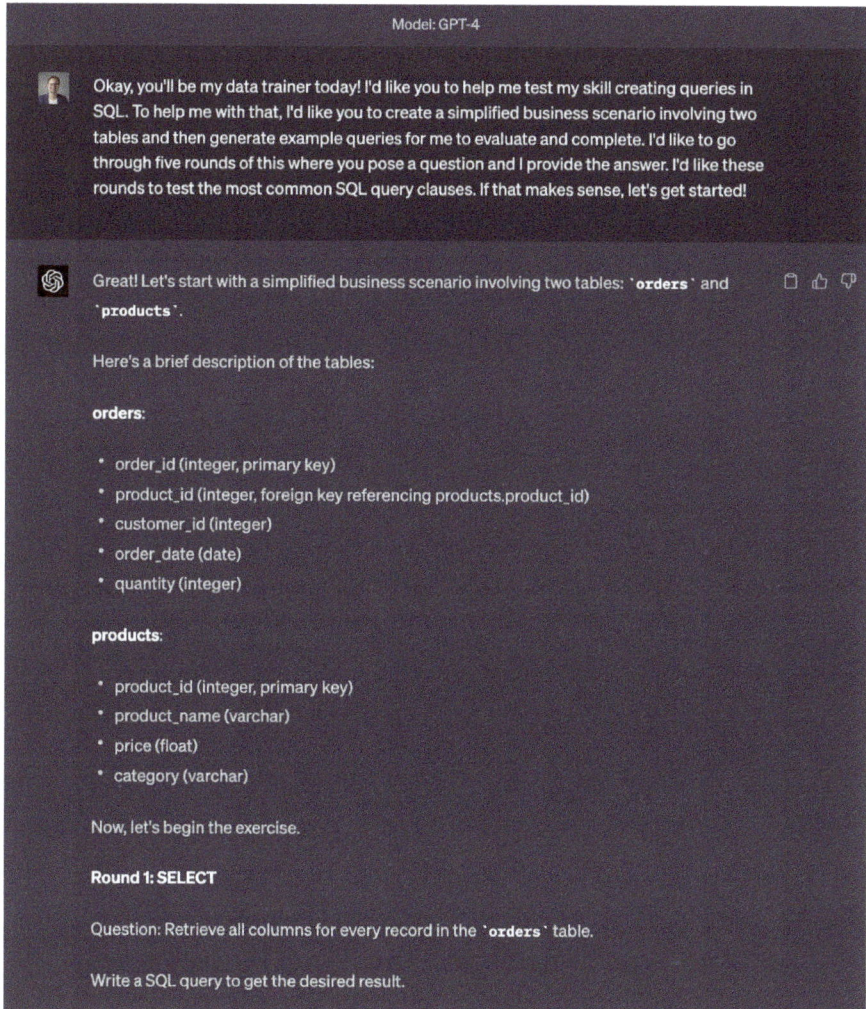

Figure 7.2. Using ChatGPT as a tutor to learn SQL

You need to be very careful about the exact AI chatbot that you have your team members use to learn data skills. Some of them, like the free version of ChatGPT, can involve data privacy concerns. Read the privacy policies carefully, and always default to leaving out company-proprietary or sensitive personal information when entering prompts into these chatbots. They can be great learning aides and guides, but there's no sense in compromising data security when using them.

Collaborate and Communicate with Data

Make it clear that you value data by the way that you ask your team to talk to each other and work with each other.

The way your team members communicate with each other is a major expression of your team's culture. A collaborative culture will be marked by open communication, sharing of information, and authenticity. A harshly competitive culture will be marked by guarded communication, withholding of information, and deceit.

This is not to say, by the way, that all internal competition is necessarily harmful. Members of high-performing teams sometimes compete in various ways. Teammates on winning sports teams, for example, compete with each other in fun ways during practice, they compete for playing time and for a starting position on the field, and they compete on the individual leaderboards. The nature of their competition, though, is constructive rather than destructive, and never sacrifices the team's success for the sake of personal gain. It's not about stepping on someone else in order to rise up. It's more about pushing each other to improve.

I really like the 68th chapter of the *Tao Te Ching* (as translated by Stephen Mitchell). It wonderfully captures this very idea of constructive competition:

The best athlete
wants his opponent at his best.
The best general
enters the mind of his enemy.
The best businessman
serves the communal good.
The best leader
follows the will of the people.

All of them embody
the virtue of non-competition.
Not that they don't love to compete,
but they do it in the spirit of play.
In this they are like children
and in harmony with the Tao.

Not everyone on your team will resonate with the idea of competition, though. And that's okay. Whether or not you're able to establish constructive, playful competition, what you want to avoid at all costs is the withholding of data and information from others on the team for competitive reasons. That is not the hallmark of a highly data literate team. A highly data literate team appreciates the importance of inclusiveness in data, and creates an open dialogue around the data so that all can share their thoughts.

A thriving data culture will go beyond mere "peppering" of data throughout conversations and meetings; communications will be steeped in data. If you imagine your team members sitting around a table, then there will be a prominent chair reserved for the data itself, which will be consulted and given a voice in every phase of every project. In your team meetings, does the data have a highly sought-after voice in the conversation, or is it totally ignored?

As we've already established, your team culture needs to stop short of overvaluing and overestimating the veracity of the voice of the data, too. Data shouldn't squash non-data voices. Your team's culture should allow for and even encourage the questioning of the data itself. If I may be so bold, I would like to present my extended version of W. Edwards Deming's quote that we considered earlier in this book:

> "In god we trust. All others must bring data. Unless there is none, in which case we go get it. Unless there's no time, in which case we apply what intuition we have. And even where there is data, we don't blindly trust it as if it were god or something."

I have to admit, my version is not nearly as pithy as Deming's original quote. And admittedly, the "or something" at the end is bordering on gratuitously verbose and totally unnecessary. That's why I'm leaving it in there. And I stand by the overall sentiment, that communicating with data is a balancing act involving inputs that are always somewhat dubious, and about which you cannot be entirely certain. I will say this, though: the reality I've laid out does make the whole affair a lot more interesting.

So the thriving data culture incorporates data into its communications, but it doesn't do so in an oppressive manner. This is true of all forms of communication, both verbal and written, and both formal and informal. Team members regularly share data, ask questions about the data, challenge it, and talk about ways to improve it.

Use Data to Improve Team Culture

Use data to better understand and improve upon the state of your own culture.

We have just considered how a team's culture can impact its usage of data, but what about the other way around? Can a team's usage of data impact its culture? More specifically, can data be used to improve a team's culture? I believe the answer is yes. While there are bound to be aspects of a culture that are difficult or even impossible to measure, much of what makes up a team's culture can be tracked and improved.

Satisfaction and Retention

First and foremost, you can collect data about how satisfied your team members are with their jobs, and how satisfied they are with the team culture. How often you collect this data is up to you, but I wouldn't recommend surveying them any more than a few times a year. Survey fatigue is real. Once or twice annually would probably suffice, as long as you regularly ask them about their satisfaction levels in less formal ways, such as in team meetings and one-on-one sessions.

When we work with our clients to find out how mature their culture is with respect to data, we recommend conducting an annual assessment. The survey itself takes between 10 and 15 minutes to complete, and it includes 50 Likert-style **scoring questions** where they get to rate the team against a variety of factors, two **multiple choice questions** where they get to select top strengths and barriers from among a list, and a single **open-ended question** that gives them a chance to share their thoughts and recommendations.

We learn a lot about their perspective and their satisfaction levels from this data, and some common pain points start to surface. A critical feature of the survey is that it's anonymous. We want to make sure team members feel comfortable saying what they think and how they feel. This kind of information is invaluable to you as a leader, because it's very actionable. It can lead directly into action plans to improve the culture in ways that you can be sure the team will appreciate. You can use an existing assessment or create your own.

Ultimately, if your team culture is great, your best team members will want to stay. Of course that's what you want. How do you measure that? You can start with **employee retention rate**, defined as the percentage of full time employees (FTE) who stayed employed from the start to the end of a particular period of time. That period of time could be a year, a quarter, or a month, depending on how granular you want it to be. The following formula tells you how to calculate that percent.

$$Employee\ Retention\ Rate\ \% = \frac{Number\ of\ employees\ who\ stayed}{Number\ of\ employees\ who\ started} \times 100$$

For example, if your team started the year with 20 people, and 2 of them left before the year was over, then your retention rate for that year would be (18/20) * 100 = 90%. It sounds simple, but it can get a bit complicated. What happens if you hire 4 people in the middle of the year? Those 4 people wouldn't factor into that year's retention rate at all. Even though you ended the year with more people (20 − 2 + 4 = 22) than you started (20), your retention rate is still 90%. Twenty started the year, and of those 20, 18 stuck around till the end.

What happens, though, if one of the 4 new hires doesn't stay for very long, and leaves the team before the year is over? Again, this wouldn't affect your retention rates at all. That's why it can be useful to measure retention on a quarterly or monthly basis, because that will catch more of these "short-timers." And it's a reason to balance the retention rate metric by also looking at **employee turnover rate**, defined as the number of people who left during a period of time divided by the average team size in that same period. Turnover rate does take into account each employee who leaves, regardless of whether they were there at the beginning.

We have just scratched the surface of the different ways some organizations track employee retention. What happens if a person gets fired or laid off—do they count? What about contractors or "gig economy" workers? The formulas can get a little complicated, but it's not rocket science. Sit down with your HR generalist and ask them how they track retention. Even if their formula is flawed in your point of view, it's probably best to use it to avoid any confusion with other leaders and teams.

Diversity and Inclusion

How diverse is your team? That can be a difficult question to answer. We tend to think about this topic in terms of gender and race, and for good reason. But the set of ways we can think about diversity is itself quite diverse. Yes, there's gender diversity, and racial and ethnic diversity. And there's also age diversity, disability and neurodiversity, geographic and linguistic diversity, and of course diversity of discipline, experience level, and background.

Another way to ask the question is this: what blind spots does our team have due to specific uniformities that exist on our team? Even if your team is very large, you'll still have major blind spots, because as humans, we tend to gravitate toward people who are like us. Because of that universal tendency, it's likely that your team members share some common attributes; it's also likely that other attributes will be underrepresented or missing entirely.

An inherent problem with blind spots is that we don't even know they're there at first. We need to become aware of them, just like when we first learned to drive, and they told us about the blind spot over our shoulder. You can use data to become aware of the attributes that are either overrepresented or underrepresented on your team. What's the makeup of your team, in terms of various characteristics? How does that makeup compare with the rest of the organization, and with the general population from which the team members were drawn? Once you identify the underrepresented or missing attributes, you can start the work of gathering the perspectives associated with those attributes.

If your team is very small—say, just a handful of people—then you'll probably have many blind spots. You may not be able to hire more people, so you have a dilemma. How can you bring a diverse perspective into the work that your team is doing? You might need to get creative, but it's not rocket science. You can research the perspectives of those whose voices are not being heard. You can form a panel of external or internal experts, an informal review board or advisory board of sorts that can weigh in on important decisions and help point out your blind spots. Yes you need to hire a diverse group of people. But you can get started today by incorporating diverse perspectives in a variety of ways.

Training and Development

The training and development opportunities that a team provides its members says a lot about the culture of that team. A team's culture should be one that encourages and enables continuous learning. If this is missing, the team environment will become stagnant very quickly.

Running a company that creates, provides, and delivers training programs, I get asked a lot how you can measure success of the training. There are, of course, a few ways to do that. Data can help you answer the following questions fairly easily:

- Utilization – sign-ups: How many people start the training?

- Utilization – completion rates: How many complete the training?

- Training time: How long does it take them to finish the training?

- Comprehension: How well do they do on objective quizzes and tests?

- Satisfaction levels: How do they rate the training?

As helpful as the answers to these questions can be, nothing compares with measuring whether some behavior changed after the training ended. What activities are they newly able to do, or what activities are they able to do in a different way? If they take a training course on data storytelling, what are their presentations or reports like after the training? Is there any evidence of improvement, and if so, how can you measure that improvement? Even if they aced the data storytelling course, and provided it with the highest possible reviews, if their presentations of data are no different after the training, then it was ineffective. Measure the training utilization rates and satisfaction levels, but don't lose sight of what really matters: the behavior change itself.

Communication and Transparency

A team with a great culture will have healthy communication and relative openness. This doesn't mean that everyone on the team knows everything all the time. That isn't always practical, and sometimes information can be sensitive. But as a leader, you need to think about what you're doing to encourage and model healthy communication. How do you use data to measure the health of your team's communication?

This is actually a big challenge for data. As I mentioned in the first chapter, I'm not a proponent of measuring things like the number of emails sent or the frequency of replies posted to group chats. You don't want to send the message that *quantity* of communication is what matters, but rather *quality* of communication.

Besides communication tool usage metrics, you can also measure meeting attendance and participation levels, you can send out surveys to your team members to gauge their perspective, and you can indirectly measure communication by measuring how well the team gets work done. I don't believe that any of these approaches is perfect, though. They each have substantial limitations and drawbacks.

This one might come down to your gut. Do you feel that bad news travels fast? In other words, when something goes wrong, do your team members let you know relatively quickly, or are you the last to find out? Do your team members trust you as the leader? How do you know? Can you think of lots of examples where your team members shared important information with each other, or is it difficult for you to think of situations where that happened?

How healthy is the level of communication between individuals or among groups? I think it's one of the most difficult things to measure with data. We can always measure activities and opinions, but just like in a marriage or a close friendship, healthy communication comes down to each person's level of comfort, and the degree to which each person is sharing the right amount of information, and doing so tactfully.

Rewards and Recognition

When you recognize your team members for their efforts and results, you should keep track of the rewards they receive. Some types of rewards will automatically be tracked in your company's financial or human resources systems. Monetary awards like raises and bonuses, for example, will automatically be tracked in your payroll system. Likewise, promotions and formal performance reviews and ratings will almost certainly be captured and retained in your company's personnel database. Sometimes, training records and completion certifications or badges are captured by a company's learning management system. Some such systems give the employee the option to put these accumulated credentials on display. Your HR generalist and your contact in the accounting department can confirm what forms of recognition are automatically tracked.

Some forms of recognition, as we discussed, will not automatically be tracked in your existing systems. Nonmonetary and informal rewards won't necessarily be logged anywhere. You can search online calendar and email histories to find records of celebratory events or written "kudos." It would be very difficult to keep track of every time you say "good job" or "thank you," nor do I think you should be that thorough about your recordkeeping.

As much as possible, use your company's formal personnel tracking system to manually log awards, rewards, and recognition that aren't already captured for you. You may need to have your HR generalist help you do this. Keeping manual lists and spreadsheets with this kind of personnel information isn't the best path. If you do so, the information is just floating around, it isn't connected to anything else, and most employees would consider this data personal and somewhat sensitive.

Celebrate and Socialize Data Successes

Take time and set aside funds to recognize and celebrate successes achieved with data.

We saved the best for last! Nothing reinforces culture quite like a good celebration. Celebrations are actually teaching moments. By throwing a fun celebration, you teach your team what you value the most, and you show them that if they deliver that value, good things will come to them.

When planning a celebration, be sure not to place additional, unwelcomed demands on your team members. "Great job working 80 hours from Monday to Friday this past week, everyone! Let's go on a full-day, mandatory, team-members-only outing on Saturday to celebrate. 'Work hard, play hard,' right?!" Imagine telling your team members' spouses and kids about those "fun" weekend plans. I've worked for more than one boss who got this wrong.

So find out what would work for your team members. Pay attention to these incidental but important details, run it by some of them to gauge their candid reaction, and if possible, schedule your celebration during normal working hours. Otherwise, let them invite their family members. While it's true that you can't make everyone happy all of the time, that's no excuse for making almost everyone unhappy. The bottom line is that you don't want your employees' spouses grumbling about you. Some of that is going to happen anyway, but you might as well not add fuel to the fire.

If you go to the trouble and expense of throwing a team celebration, it seems like it would be a huge missed opportunity if people on your team didn't know exactly what they were celebrating, or how the success was achieved. If all they know is that the team hit some lofty goal or objective but they don't know who did what, then all they'll learn is that the company has extra funds to blow on extravagant parties. I bet you can guess what they'll wonder next.

My recommendation is to do more than *celebrate* wins: you need to *socialize* wins. How do you do that? Have the people most responsible for the achievement present what they did, how they did it, what they learned, and how those learnings can transfer across the team and organization. They don't have to write a book about their project or make a 50-slide PowerPoint presentation about it. Just a quick email with key points, or a 10-minute time slot in the next team meeting to do some show-and-tell would suffice. The point here isn't to reward people who succeed by giving them extra work. You aren't asking for something polished or extravagant from them. But give them a chance to "toot their horn." And challenge everyone else on the team to find ways to transfer the parts that could work well for them, too.

How does this apply to data? Well, if you want to stress the importance of data in your culture, ask the people sharing their win to talk about how they leveraged data to achieve success. If you really want to get the message across, consider asking a team that *didn't* succeed, but still did amazing things with data in spite of the less-than-stellar outcome. Socialize winning approaches, methods, and techniques, not just winning outcomes.

Culture Summary

There's no denying that culture is critical. No one has more influence on the team's culture than you, as your team's leader. If you are negative and critical, then your team's culture will become caustic. If you are positive and motivational, then your team's culture will become vivid and alive.

Don't mistake this to mean that you have to have a bubbly personality, by the way. I have seen very reserved and almost subdued leaders create a positive team culture, and I have seen very energetic, cheerleader types create a cultural nightmare. It isn't about the style of your leadership as much as it's about the substance behind that style.

Care about them as individuals, care about the work you're doing together, and put your ego aside. Be willing to listen, and be authentic. Back up what you say with real actions that match. When it comes to data, embed it right in the heart of your team's culture.

We considered seven different tactics to implement when creating a thriving culture as a leader. We can restate these seven tactics as seven powerful questions as follows:

1. What data communities or groups can your team members join?

2. To what degree is data a part of your team's identity and purpose?

3. How are you recognizing and rewarding the successes they have with data?

4. What data training and development opportunities are you providing for them?

5. Is data incorporated into the way you communicate and collaborate?

6. How are you using data to improve the state of your culture?

7. What are you doing to celebrate and socialize your team's "data wins?"

CONCLUSION

With that, you have completed your tour through the seven categories of data leadership! Congratulations on making it all the way through, and thank you for joining me and letting me be your tour guide. I hope that I was able to give you some helpful pointers and open your eyes to some new opportunities to make an impact.

I hope that at this point, you appreciate that being a great leader in the age of data involves more than just giving your team members access to clean **data** and powerful **technologies**. You need to provide them with those things, but you also need to establish sound principles of **ethics**, connect the data to your team's **purpose**, make sure the **people** on your team have the knowledge and skills they need, design your team's **processes** to leverage data, and create a **culture** that promotes fluency in a shared language.

That's a lot to juggle, isn't it? In the age of data, the role of the leader is not for the faint of heart, at least not if you want to become a really great one. The best approach is to consider each of these aspects of leadership, and then decide what to focus on, in what order, and why. These seven categories can be a framework within which you can strategically pick and choose your battles so that you can win the overall war instead of losing it. It's possible, by the way, to win battles in such a way that you lose the war. So you will have to choose wisely.

But what, exactly, is a wise approach? Well, as you look back on these seven categories, start by assessing which are your strong suits and which are your weak ones. Do your peers and direct reports agree with your own assessment of the current state of your leadership abilities? If not, then in what ways do your opinions differ substantially? If your own leadership abilities were the only consideration, then you'd be done. You'd simply need to start at the bottom of the list and work your way up, improving in your weakest categories one by one.

But that's only half of the equation. The other half deals with the relative importance of each category in your team's current environment, as well as logical dependencies between them. For example, if you struggle with data ethics, you'll want to resolve that right away. In fact, I wouldn't recommend you train anyone or purchase any fancy data tools if that cornerstone isn't in place. The world doesn't need any more unethical data teams.

Similarly, some teams may place a higher premium on processes than others. This may simply be due to the nature of their work. If your team spends a high percentage of their time running and rerunning through recurring processes, then it may be critical for you to invest time and energy improving that aspect of your leadership. For other teams, process may be less important than, say, culture. Perhaps a team has just had layoffs and morale is very low. In such a situation, really acing culture and starting to gain some momentum that way might need to take precedence over a continuous improvement initiative to perfect your team's processes.

This is not an exact science. There's no formula that will apply to each and every situation. The art of being a great leader involves understanding your own strengths and weaknesses while simultaneously intuiting the importance associated with each. Only you can commit to a particular game plan. And like any plan, you'll need to constantly adjust it. Sometimes you will need to make major adjustments. As we've been reminded lately, situations can drastically change overnight, and the development path you were on may no longer make any sense at all. Other times, even a small tweak to your approach can make a massive difference in the impact you're able to have on your team.

I wish you all the best on your leadership journey! The world is changing and so are you. The importance of data in the world is only increasing, and your abilities are increasing, too. There is a path forward in which you become a great leader in this present age of data. It's up to you to develop a keen vision of that future, and then go make it happen.